An Introduction to the Engineering of Fast Nuclear Reactors

This book is a resource for both graduate-level engineering students and practicing nuclear engineers who want to expand their knowledge of fast nuclear reactors, the reactors of the future. The book is a concise yet comprehensive introduction to all aspects of fast reactor engineering. It covers topics including neutron physics, neutron flux spectra, Doppler and coolant temperature coefficients, the performance of ceramic and metal fuels under irradiation, the effects of irradiation and corrosion on structural materials, heat transfer in the reactor core and its effect on core design, coolants including sodium and lead-bismuth alloy, coolant circuits, pumps, heat exchangers and steam generators, and plant control. The final chapter covers all aspects of safety including operational safety and hypothetical accidents. The book includes discussions of gas coolants, the use of reactors to consume radioactive waste, and accelerator-driven subcritical systems.

Anthony M. Judd has more than 40 years experience in nuclear engineering, including managing the operation of a fast reactor power station and a period as Chief Technologist for Fast Reactors during which he was responsible for the entire UK national fast reactor R&D program. He spent periods at the Argonne National Laboratory (United States) and at the University of Cambridge Engineering Department. In addition to his research publications and presentations, he is the author of *Fast Breeder Reactors: An Engineering Introduction*.

AN INTRODUCTION
TO THE ENGINEERING
OF FAST NUCLEAR REACTORS

Anthony M. Judd

CAMBRIDGE
UNIVERSITY PRESS

CAMBRIDGE
UNIVERSITY PRESS

Shaftesbury Road, Cambridge CB2 8EA, United Kingdom

One Liberty Plaza, 20th Floor, New York, NY 10006, USA

477 Williamstown Road, Port Melbourne, VIC 3207, Australia

314–321, 3rd Floor, Plot 3, Splendor Forum, Jasola District Centre, New Delhi – 110025, India

103 Penang Road, #05–06/07, Visioncrest Commercial, Singapore 238467

Cambridge University Press is part of Cambridge University Press & Assessment, a department of the University of Cambridge.

We share the University's mission to contribute to society through the pursuit of education, learning and research at the highest international levels of excellence.

www.cambridge.org
Information on this title: www.cambridge.org/9781107034648

First published 2014

A catalogue record for this publication is available from the British Library

Library of Congress Cataloging-in-Publication data
Judd, A. M.
An introduction to the engineering of fast nuclear reactors / Anthony M. Judd.
    pages cm.
"A revision and extension of Fast breeder reactors : an engineering introduction, published in 1981" – Preface.
Includes bibliographical references and index.
ISBN 978-1-107-03464-8 (hardback)
1. Fast reactors. 2. Breeder reactors. 3. Nuclear engineering. I. Judd, A. M. Fast breeder reactors. II. Title.
TK9203.F3J84   2014
621.48′34–dc23   2013030947

ISBN   978-1-107-03464-8   Hardback

CONTENTS

PREFACE

This book is a revision and extension of *Fast Breeder Reactors: An Engineering Introduction*, published in 1981. I have rewritten much of it in the light of developments in fast reactor technology that have taken place in the subsequent three decades, and to take account of the new applications for fast reactors that have been suggested.

It is intended for the newcomer to the study of fast reactors, either as a student or at a later stage of his or her career. It will probably be most useful to someone who already has some knowledge of nuclear reactors. There are many excellent introductory texts for the beginner in nuclear engineering but they all concentrate on thermal reactors. The purpose of this book is to provide an up-to-date account of fast reactors for those who want to take the next step.

Fast reactor technology has become a wide field, so wide that it is not possible to cover all of it in depth in a single book of reasonable length. What I have attempted is to cover the whole in sufficient detail to allow the reader to understand the important features, and to provide suitable references for further study. I have gone into detail on the neutron physics because any fast reactor engineer, whether he or she is a designer, an operator or a researcher, needs to understand how the machinery works at a basic level. I have also attempted to include the results of experience, often hard-won, of operating a fast reactor power station.

I have divided the subject matter up in chapters according to discipline. Chapter 1 about the physics of fast reactors is the most detailed and mathematical. This is to give those who have to use the numbers produced by the complex computer codes that predict reactor performance some idea of where they come from. Chapter 2 is mainly about the chemistry of fast reactor fuel. Chapters 3 and 4 are about the application of mainly conventional engineering disciplines to fast reactors so they contain less theoretical detail. In Chapter 5 I have tried to show how safety can be attained by careful attention to detail in design. The Introduction includes an explanation of the difference between fast reactors and thermal reactors and a brief summary of the history of fast reactor development.

I wish to thank Argonne National Laboratory for permission to reproduce Figures 2.19, 2.22 and 2.25.

Many of my colleagues in the atomic energy industry have been very generous in helping me to write this book and its predecessor. They are for too numerous to mention by name. By way of thanks I wish to dedicate this account of the technology to the hundreds of engineers, scientists and technicians whose achievements made possible the success of the British Fast Reactor project, started in 1946 and abandoned prematurely in 1993.

Anthony M. Judd

INTRODUCTION

WHAT FAST REACTORS CAN DO

Chain Reactions

Early in 1939 Meitner and Frisch suggested that the correct interpretation of the results observed when uranium is bombarded with neutrons is that the uranium nuclei undergo fission. Within a few months two very important things became clear: that fission releases a large amount of energy, and that fission of a nucleus by one neutron liberates usually two or three new neutrons. These discoveries immediately disclosed the possibility of a chain reaction that would produce power.

There was a difficulty, however, in making a chain reaction work. Natural uranium consists of two isotopes: ^{235}U (with an abundance of 0.7%) and ^{238}U (99.3%). Of the two only ^{235}U is "fissile", meaning that fission can be induced in it by neutrons of any energy. On the other hand ^{238}U undergoes fission only if the neutrons have an energy greater than about 1.5 MeV, and even then they are more likely to be captured or scattered inelastically.

Figure 1 shows the fission cross-sections of ^{235}U and ^{238}U and the capture cross-section of ^{238}U as functions of neutron energy. Because ^{238}U is so abundant in natural uranium capture in it dominates over fission in ^{235}U except at energies below about 1 eV.

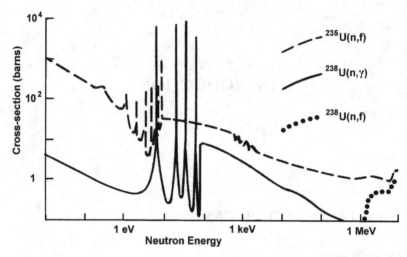

Figure 1 Cross-sections of ^{235}U and ^{238}U.

Neutrons generated in fission have average energies of about 2 MeV and at that energy cannot sustain a chain reaction in natural uranium. If a neutron survives many scattering interactions, however, its kinetic energy decreases until it is in thermal equilibrium with the atoms by which it is being scattered. It is then known as a "thermal" neutron and its most probable energy is about 0.025 eV.

If a chain reaction is to take place, therefore, either the fission neutrons have to be reduced in energy to near the thermal level, in which case natural uranium can be used, or the proportion of ^{235}U has to be increased substantially. Both of these routes were followed in the early work on nuclear reactors. The first led to the development of "thermal" reactors and the second to "fast" reactors, so called because the neutrons causing fission are fast as opposed to thermal.

Breeding and Consumption

It is possible to make use of the neutrons that are not needed to maintain the chain reaction in various ways. The most important is for breeding. When a neutron is captured in ^{238}U the ^{239}U that is formed decays in the following way:

$$^{239}\text{U} \xrightarrow[\text{23.5 min}]{\beta-} \rightarrow {}^{239}\text{Np} \xrightarrow[\text{2.35d}]{\beta-} \rightarrow {}^{239}\text{Pu} \xrightarrow[\text{24360y}]{\alpha} \rightarrow {}^{235}\text{U} \text{ etc.}$$

The times shown are the half-lives for the decay processes. As far as reactor operation is concerned the long-lived plutonium isotope ^{239}Pu is the end-product of the chain.

^{239}Pu has nuclear properties quite similar to those of ^{235}U and it can be fissioned by neutrons of all energies. Neutron capture thus provides a route for converting ^{238}U into fissile material, so ^{238}U is called a "fertile" isotope. ^{232}Th, which is the only naturally occurring isotope of thorium, is also fertile. It behaves very similarly to ^{238}U: the ^{233}Th formed on capture of a neutron decays in a chain to ^{233}U which is long-lived and fissile.

$$^{232}\text{Th} \xrightarrow[\text{22.1 min}]{\beta-} \rightarrow {}^{233}\text{Pa} \xrightarrow[\text{27.4d}]{\beta-} \rightarrow {}^{233}\text{U} \xrightarrow[\text{162000y}]{\alpha} \rightarrow {}^{229}\text{Th} \text{ etc.}$$

Thus there are two naturally occurring fertile isotopes, ^{232}Th and ^{238}U, and three related fissile isotopes: ^{233}U, ^{235}U and ^{239}Pu. There are other fissile and fertile isotopes but these five are the most important.

This ability to convert fertile isotopes to fissile raises the possibility of "breeding" new fissile material, but this can be done only if enough neutrons are available. The average number of neutrons liberated in a fission is denoted by $\bar{\nu}$. Its value depends on which isotope is being fissioned and on the energy of the neutron causing the fission, but in most cases it is about 2.5. We have seen that the fact that $\bar{\nu}$ is greater than 1 makes a chain reaction possible: the fact that it is greater than 2 is almost equally important. If we have a reactor in which on average one neutron from each fission causes another fission to maintain the chain reaction, and if in addition more than one of the other neutrons is captured in fertile material, then the total number of fissile nuclei will increase as the reactor operates. Such a reactor is called a "breeder".

It is sometimes said that a breeder reactor generates more fuel than it consumes. This is rather misleading. The reactor produces more fissile material than it consumes, but to do this it depends on a supply of fuel in the form of fertile material.

Although $\bar{\nu} > 2$ suggests the possibility of a breeder reactor the requirement for breeding to take place is more complicated. When a neutron interacts with a fissile nucleus it does not necessarily cause fission. It may be captured, and if it is, it is effectively lost. The important quantity in determining whether breeding is possible is the average number of neutrons generated per neutron absorbed. This is denoted by η, where

$$\eta = \bar{\nu}\sigma_f/(\sigma_f + \sigma_c).$$

η, which is sometimes called the "reproduction factor", is a function of the neutron energy E, and its variation with E for the three fissile isotopes is shown in Figure 2.

Of these η neutrons one is needed to maintain the chain reaction. Some of the remainder are lost either because they diffuse out of the reactor or because they are captured by some of the other materials present, such as the coolant or the reactor structure. The others are available to be captured by fertile nuclei to create fissile nuclei. If we denote the number of neutrons lost per neutron absorbed in fissile material by L and the number captured in fertile material by C, then C is the number of fissile nuclei produced per fissile nucleus destroyed and is given by

$$C \simeq \eta - 1 - L$$

(This is only a rough value because there are other things that may happen to neutrons).

If C is greater than one, as it must be if the reactor is to breed, it is known as the "breeding ratio". If it is less than one it is called the "conversion ratio". There is no logical reason for the existence of two names for C. The usage grew up because different words were used in the contexts of different reactor systems.

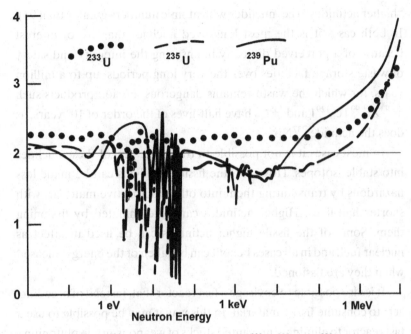

Figure 2 The reproduction factor, η, for ^{233}U, ^{235}U and ^{239}Pu.

In practice L cannot be reduced below about 0.2, so that breeding is possible only if η is greater than about 2.2. Figure 2 shows how this can be brought about. A fast reactor using any of the three fissile materials can be made to breed, although ^{239}Pu gives the widest margin and ^{235}U will allow breeding only if the energy of the neutrons causing fission is not allowed to fall much below 1 MeV. In all cases the higher the neutron energy the better the breeding ratio. A ^{233}U-fuelled thermal reactor is just able to breed but the margin is very slender. The most widely favoured breeder system is based on the use of ^{238}U and ^{239}Pu in fast reactors, but there is also a certain amount of interest in the ^{232}Th – ^{233}U system, also in fast reactors.

A fast reactor does not necessarily have to be a breeder. The excess neutrons can be used in other ways. One such is to use them to consume radioactive waste materials by transmutation. This process can be applied to two classes of radioactive waste: fission products and "higher actinides" (i.e. nuclides with atomic numbers greater than 94). In both cases it is the most long-lived nuclides that are of interest because of a perceived difficulty in ensuring the integrity and safety of waste storage facilities over the very long periods, up to a million years, for which the waste remains dangerous. Fission products such as ^{93}Zr, ^{99}Tc, ^{129}I and ^{135}Cs have half-lives of the order of 10^6 years, as does the actinide ^{237}Np.

In most cases it is not possible to transmute any of these nuclides into stable isotopes. However, the fission products can be made less hazardous by transmuting them into other radioactive materials with shorter half-lives. Higher actinides can be eliminated by fissioning them. Some of the fissile higher actinides can be used in effect as nuclear fuel, and in all cases benefit can be taken of the energy released when they are fissioned.

It is also possible to envisage a fast reactor that, instead of breeding, acts to consume fissile material. In this way it may be possible to use a fast reactor to eliminate unwanted stocks of weapons-grade plutonium.

Energy Resources

Even if fast reactors are used to consume radioactive waste in due course their main function is likely to be to breed fissile material because in this way they can have a transforming effect on the world's energy resources.

Consider a uranium-fuelled reactor in which N atoms of ^{235}U are fissioned. While this is happening CN new fissile atoms of ^{239}Pu can be produced. If these in turn are fissioned in the same reactor and the conversion or breeding ratio C is unchanged (this is unlikely to be quite true because the fissile material has been changed, but the effect on the argument is not important), a further C^2N fissile atoms are produced. If these are fissioned, C^3N are produced, and so on indefinitely. The total number of atoms fissioned is therefore $N(1 + C + C^2 + \ldots)$. If $C < 1$, the series converges and its sum is $N/(1 - C)$.

Conversion ratios for ^{235}U-fuelled thermal reactors are in the range 0.6 (for light-water reactors) to 0.8 (for heavy-water reactors and gas-cooled reactors). L is particularly large in light-water reactors because neutrons are readily absorbed by hydrogen.

If the fuel is natural uranium N cannot exceed 0.7% of the total number of uranium atoms supplied. If the reactor is a thermal reactor with a conversion ratio of 0.7 and the plutonium bred is recycled indefinitely the total number of atoms fissioned cannot exceed $0.7/(1-0.7) \approx 2.3\%$ of the number of uranium atoms supplied.

In a real system not even this number can be fissioned. When the fuel is reprocessed to remove the fission products and the excess ^{238}U some ^{235}U is inevitably lost. In addition some ^{239}Pu is lost by conversion to higher isotopes of plutonium. As a result thermal converter reactors can make use of at most about 2% of natural uranium.

For a breeder reactor, however, with $C > 1$, the series diverges and in principle all the fertile atoms supplied can be fissioned. In practice, however, some are lost for the reasons mentioned earlier and the limit is around 60% of the fertile feed. Thus from a given quantity of natural uranium fast breeder reactors can fission about 30 times as many atoms as thermal converters and as a result can extract about 30 times as much energy.

To determine the importance of this difference we have to know how much uranium and thorium are available. The amount depends on the price, and a 2010 estimate by the World Energy Council suggests that, worldwide, about 230000 tonnes of uranium are recoverable at a price up to $40/kg, but that if the price were to rise to $260/kg ten times as much would be accessible. The extent of reserves of thorium is much less certain but seems to be comparable with those of uranium. Thorium can be made available as an energy resource only by means of breeder reactors.

Complete fission of a tonne of uranium, were that possible, would generate about 1 TWd, or 0.09 EJ, of energy in the form of heat. (An exajoule, EJ, is 10^{18} joules.) Thus if all the $40/kg uranium in the world were used as fuel for thermal reactors that, with recycling, fissioned 2% of the feed, some 400 EJ thermal would be produced. If the same uranium were to be recycled to exhaustion in fast breeder reactors it would produce about 12000 EJ. But if the higher utilisation would allow the higher price of \$260/kg to be paid so that the greater resource became available the production would rise to 1.2×10^5 EJ. These quantities can be compared with about 9.0×10^{11} tonnes of "proved recoverable" coal reserves that could yield some 3000 EJ, or 1.6×10^{11} tonnes of "proved recoverable" oil that could yield about 800 EJ. In 2007 some 71 EJ of electricity was generated throughout the world.

There is considerable uncertainty about the true extent of mineral reserves in the earth's crust because new discoveries continue to be made. However, in spite of this the overall conclusion is that uranium used in thermal reactors has the potential to make a contribution to the world's energy consumption that is comparable with, but smaller than, that of oil, whereas uranium used in fast breeder reactors could contribute considerably more than, possibly 40 times as much as, all the world's fossil fuel. Thorium used in breeder reactors could probably make a similar contribution. Together they could provide the world with all the energy it needs for centuries to come. And they would do this without adding to the amount of carbon dioxide in the atmosphere.

HOW FAST REACTORS HAVE BEEN DEVELOPED

The Early Years

The history of fast breeder reactors is quite dissimilar from that of thermal reactors. From the earliest days after the Second World War the development of different types of thermal reactor was pursued in different countries: light-water reactors in the United States, heavy-water reactors in Canada and gas-cooled reactors in the United

Kingdom, for example. Only towards the end of the 20th century did the various nationally based lines of development converge.

In contrast virtually the same path was followed in all the countries where work on fast reactors was done. The reason for this seems to have been that until the 1960s at least fast reactors were seen to be commercially valuable only well into the future, so that the advantages of cooperation appeared to outweigh the disadvantages of aiding possible competitors. Thermal reactors on the other hand were commercially important from the start and were developed in competition, which restricted the exchange of ideas and allowed different concepts to flourish.

International cooperation played a major role in fast reactor development for two main reasons. Firstly the nuclear data on which designs had to be based were inadequate until the 1960s. There was a lot to be gained from worldwide cooperation in measuring neutron cross-sections to the required accuracy and exchanging and comparing the results. Secondly cooperation to ensure the safety of fast reactors was desirable even when there was competition in other areas.

This need to exchange information resulted in, among other things, a series of international conferences on fast reactors that were addressed mainly to the problems of reactor physics and safety and were held in the United States and various European countries throughout the 1960s and 1970s. These, together with the continual publication of information in the scientific press, kept the thinking in different countries from diverging and encouraged parallel development.

In one respect it is not altogether certain that this was an advantage. The use of liquid metals as coolants acquired a great deal of momentum, mainly because "everyone did it", and the search for alternatives was discouraged. Gas has certain advantages as a coolant, but at the time of writing no group has been able to develop a gas-cooled fast reactor to the point where it can be assessed fairly in comparison with a liquid-metal-cooled reactor.

The Era of Metal Fuel

Before about 1960 it was thought that a high breeding ratio was the most important quality of a fast reactor. To achieve this the mean energy of the neutrons has to be kept high, and this requires that extraneous materials, especially moderators, should be excluded as far as possible from the reactor core. As a result the early reactors had metal fuel, the metal being either enriched uranium or plutonium, alloyed in some cases with molybdenum to stabilise it to allow operation at higher temperature.

The critical masses of these reactors were small and the cores were correspondingly small so that for high-power operation they had to be cooled by a high-density coolant (to avoid impossibly high coolant velocities). Hydrogenous substances were precluded because hydrogen is a moderator, so liquid metals were used. In most cases the coolant was sodium or sodium-potassium alloy. Some early experimental reactors were cooled with mercury but this fell out of favour because of its toxicity, cost, and low boiling point.

The many neutrons that leaked from the small cores of these reactors were absorbed in surrounding regions of natural or depleted uranium where the majority of the breeding took place. These were known as breeders, or blankets.

The first generation of low-power experimental fast reactors were built in the late 1940s and early 1950s to demonstrate the principle of breeding and to obtain nuclear data. They included CLEMENTINE and EBR-I in the United States, BR-1 and BR-2 in the Union of Soviet Socialist Republics, and ZEPHYR and ZEUS in the United Kingdom. CLEMENTINE, ZEPHYR, and BR-1 and 2 used plutonium fuel, which in the early years was more readily available than highly enriched uranium. Apart from ZEUS, which was a zero-power mock-up of the later DFR, they all had very small cores, the largest being EBR-I (6 litres), which was a small power reactor with an output of 1.2 MW.

When it came to higher powers, however, the volume of the core had to be increased to keep the heat fluxes down to reasonable levels and to allow for the extra coolant flow. The result was EBR-II and EFFBR (the Enrico Fermi Fast Breeder Reactor) in the United States, and DFR (the Dounreay Fast Reactor) in the United Kingdom. When they were designed they were seen as prototypes of the reactors to be used in power stations, but as they were built it began to be recognised that they would be the end-point of the development of metal-fuelled fast reactors, and the principal use to which DFR and, for many years, EBR-II were put was to test oxide fuel for the next generation.

The Importance of Burnup

Around 1960 it became clear that there is more to a profitable fast reactor than a high breeding ratio. The fuel itself is expensive because of the original cost of the fissile material and the cost of fabricating it into fuel elements and reprocessing it after it has been irradiated in the reactor.

Fuel cannot remain in the reactor core indefinitely for a number of reasons. As irradiation proceeds the fissile material is used up. This is offset to some extent by the breeding of new fissile material but as most breeding takes place in the breeder rather than the core there is a net loss of fissile material from the core and the reactivity declines. At the same time fission products are formed. These have a small effect on reactivity in a fast reactor (unlike a thermal reactor in which they absorb the low-energy neutrons), but they disrupt the crystal structure of the fuel material and cause it to swell, and they corrode the cladding material. In addition, the cladding is weakened by the fast neutrons so that in the end it loses its integrity and allows radioactive material to escape into the coolant.

When the fuel has been irradiated in the reactor for as long as it can stand it has to be removed and stored for a period to allow the most intense fission-product radioactivity to decay. It is then reprocessed chemically to remove the fission products and replace the fissile material, refashioned into fresh fuel elements, and returned to the reactor to be used again. The more frequently this has to be done the greater the cost, both because more reprocessing has to be paid for and also because more of the expensive fuel is lying unused waiting for reprocessing or being reprocessed.

The amount of irradiation the fuel can stand before it has to be removed and reprocessed is known as the "burnup". It can be measured by determining either the fraction of the total number of atoms of uranium and plutonium (or thorium and uranium if that cycle is being used) that are fissioned, or by determining the total amount of heat transferred from the fuel. The two are equivalent because one fission releases approximately the same amount of energy whatever isotope is fissioned, and because whatever its chemical form the mass of the fuel material is almost entirely made up by the mass of the uranium, plutonium or thorium atoms. It so happens that fission of all the uranium and plutonium (i.e. all the "heavy atoms") would, if it were possible, release about 10^6 MW days per tonne of fuel.

Experience with the early reactors showed that metal fuel can stand no more than about 1% burnup, or 10,000 MWd/tonne, whereas fuel in the form of dioxides (either UO_2, PuO_2 or a mixture of the two) can stand much higher burnup, up to 10% or more. It became obvious that the reprocessing and inventory costs of fuel that could stand only 1% burnup would be prohibitively high. Metal fuel had another disadvantage in that it cannot tolerate operation at high temperature. Phase changes in the crystal structure of the metal itself, together with the difficulty of finding a cladding material compatible with both fuel and coolant, limited temperatures to about 250 °C. This severely limited the thermodynamic efficiency in converting heat to work and so restricted the electrical power output.

For these reasons, in most quarters, metal fuel was rejected in favour of oxide fuel. The first power demonstration reactors in France (Rapsodie) and Russia (BOR-60), both built towards the end of the 1960s, were oxide-fuelled.

Oxide Fuel and Sodium Coolant

Oxide, however, is far from perfect as a fuel material. The oxygen acts as a partial moderator, reducing the mean neutron energy and therefore decreasing the breeding ratio. The low thermal conductivity of the oxide is another disadvantage because it means that the fuel elements have to be very slender and the cost of manufacturing them is high.

There are on the other hand compensating advantages apart from the ability of oxide to withstand high burnup. The low mean neutron energy means that there are enough neutrons in the ^{238}U resonance region to make the Doppler effect important. This results in a prompt negative temperature coefficient of reactivity, which in some cases offsets the positive sodium temperature coefficient due to the large size of the core of a high-power reactor. These effects are described in Chapter 1.

Oxide fuel can be operated at a higher temperature than metal, and it can be clad in stainless steel, which is cheaper than the refractory metals used in metal-fuelled reactors. Above all there is a lot of experience about it because it is widely used as a fuel for thermal reactors.

This last is the reason why oxide is preferred to carbide. A mixture of UC and PuC would have important advantages (higher thermal conductivity, higher density, less moderating effect), but because it was not so well understood caution dictated that oxide should be preferred. It is possible that carbide fuels will be used widely at some time in the future.

Other fuels have been tried. Experiments were done on cermets, consisting of a sintered mixture of stainless steel and oxide powders. This had the advantage of high thermal conductivity and a porous structure that can accommodate fission products. It was rejected because absorption of neutrons in the steel was very detrimental to the breeding ratio. Molten fuel was tried in the LAMPRE experiment. This reactor had molten plutonium fuel clad in tantalum. The problems of corrosion and of the accumulation of bubbles of fission-product gas in the molten fuel were so severe that this line of development was not pursued.

So by 1970 mixed oxide fuel, stainless steel cladding and structure, and sodium coolant became accepted almost universally as the route for the development of fast breeder reactors. These materials restrict the designers' choice of variables such as the dimensions of the fuel elements and the core and as a result all fast breeder reactors of the period, from whatever country, showed marked similarities. The British Prototype Fast Reactor (PFR), French Phénix, Russian BN-350, German SNR-300, Japanese Monju, and the proposed Clinch River Breeder Reactor (CRBR) in the United States were, as far as the design of the reactor core is concerned, very much alike. They were all prototypes intended to be followed by full-scale production reactors for commercial generation of electricity such as Super-Phénix in France, the Commercial Fast Reactor (CFR) in Britain, and BN-600 in Russia.

The Period of Decline

In the last quarter of the 20th century the development of fast reactors declined. Nuclear power in general was set back, partly because of the Chernobyl accident, so that the fear that supplies of uranium would run out receded and with it the perception that breeding was neces-sary, at least for many decades. In addition, in the West in particular the public mood turned away from nuclear power and also in many countries from large government-funded development projects. The world's largest fast reactor power station, Super-Phénix, was built in France but suffered a series of setbacks so that it lost political sup-port and was shut down prematurely. In Britain PFR struggled with a series of technical problems, and there was a sodium fire at Monju in Japan. A European collaboration succeeded in designing a "next-generation" fast reactor EFR (the European Fast Reactor) but there was no interest in constructing it. In the United States the development of oxide-fuelled reactors came to an almost complete halt. Only BN-600 in the Soviet Union and then Russia was conspicuously successful as a reliable power station.

Amid these setbacks, however, there were several signs of potential for the future. A widespread concern about the disposal of nuclear waste gave rise to interest in the use of fast reactors to consume, or "incinerate", hazardous radionuclides. A concern about safety led to the suggestion that subcritical reactors driven by particle accelerators would be less prone to damaging reactivity accidents. Experience of sodium fires led to reexamination of alternative coolants such as helium or lead. The latter was given impetus by the release of information about the Soviet submarines powered by small lead-cooled fast reactors that had been developed, unknown to the rest of the world, in the 1970s. In the United States development of metal fuel continued in the 1980s and early 1990s. EBR-II was used to demonstrate burnup of nearly 20%, and this, coupled with work on pyro-chemical reprocessing of the fuel, led to a proposal for an "integral fast reactor" (IFR) system.

Throughout this period information was exchanged under the aegis of the International Atomic Energy Agency. The IWGFR (International Working Group on Fast Reactors) was established in 1967, held regular technical discussion meetings and issued several reports including a series of "Status Reports" on fast reactor development worldwide.

The 21st Century

The decline in Europe and America did not affect Asia. A long-standing programme in India had continued, somewhat separated from work in the rest of the world. In Japan activity had continued in spite of the Monju fire, and there was continuing interest in the Republic of Korea. Most significantly work had started in China. Nuclear power is seen to be necessary for the future economic health of all of these countries and breeding will be an essential component by the second half of the century. Because there is so much more experience with it than with any alternative the main emphasis is on sodium as a coolant.

In the West as well the revival of nuclear power has been accompanied by increased interest in fast reactors, but here it is not so clear that the future lies with the sodium-cooled oxide-fuelled critical breeder that had been the norm in the 1970s. Fast reactors may have diverse applications (consumption as well as breeding) in the years to come, and there may be several design variations (lead or gas coolants as well as sodium, accelerator-driven subcritical reactors as well as critical, and metal fuel as well as oxide).

Because there is so much more experience of sodium-cooled breeders most of the content of the following chapters is about systems of this type, but the intention is also to provide an introduction to the alternative coolants, fuel materials and design styles that may become important as fast reactors are deployed for a range of purposes in a growing and diversifying nuclear industry across the world.

GENERAL REFERENCES

Chang, Y. I. and C. E. Till (2011) *Plentiful Energy: The Story of the Integral Fast Reactor*, CreateSpace online publishers

Forrest, J. S. (Ed.) (1977) *The Breeder Reactor*, Scottish Academic Press, Edinburgh

IAEA (2013) *Status of Fast Reactor Research and Technology Development* Technical Report TECDOC-1691, International Atomic Energy Agency, Vienna

IWGFR (1985) *Status of Liquid Metal Cooled Fast Breeder Reactors* Technical Report 246, International Atomic Energy Agency, Vienna

IWGFR (1999) *Status of Liquid Metal Cooled Fast Reactor Technology* Technical Report TECDOC-1083, International Atomic Energy Agency, Vienna

IWGFR (2007) *Liquid Metal Cooled Reactors: Experience in Design and Operation* Technical Report TECDOC-1569, International Atomic Energy Agency, Vienna

Judd, A. M. (1981) *Fast Breeder Reactors: An Engineering Introduction*, Pergamon, Oxford

Judd, A. M. (1983) Fast Reactors, pp 297–333 in W. Marshall (Ed.) *Nuclear Power Technology, Volume 1: Reactor Technology*, Clarendon, Oxford

Sweet, C. (Ed.) (1980) *The Fast Breeder Reactor: Need? Cost? Risk?*, Macmillan, London

Waltar, A. E. and A. B. Reynolds (1981) *Fast Breeder Reactors*, Pergamon, New York

Waltar, A. E., D. R. Todd and P. V. Tsvetkov (2012) *Fast Spectrum Reactors*, Springer, New York

1

PHYSICS

1.1 INTRODUCTION

1.1.1 Physics and Design

Whether the purpose of a fast reactor is to generate power, to breed fissile material, to consume fissile material or to consume nuclear waste products, whether its chain reaction is to be critical and self-sustaining or subcritical and driven by an external source of neutrons, reactor physics – the understanding of the nuclear reactions that take place in it – is fundamental to its design in two ways. Firstly, criticality is a question of reactor physics. The designer of the reactor has to determine the size and composition needed to make the reactor critical or to achieve the required degree of subcriticality, to predict the effect on reactivity of movement of the control rods and the burnup of the fuel, and to estimate the reactivity changes that come about in the course of normal operation and under abnormal conditions. Secondly, he or she has to know the rate at which various nuclear reactions take place, for on these depend the power generated and its distribution within the reactor, the burnup of the fuel, the breeding or destruction of fissile material and nuclear waste, the alteration of the properties of the materials of which the reactor is constructed, the build-up of radioactivity, and the need for radiation shielding.

Reactor physics is not, however, the only important influence on design. Heat transfer, structural, metallurgical, and safety considerations are also important, and the design ultimately chosen is a compromise. In reaching this compromise a designer's overriding aim is that the reactor should be as effective as possible in achieving its objectives, provided that it is safe.

1.1.2 Comparison with Thermal Reactors

The physics of fast reactors differs considerably from that of thermal reactors. The most important difference is that the composition of the fuel is different. In a fast power reactor the fraction of fissile material in the fuel is about 20–30% compared with 0.7–3% in a thermal reactor. In a reactor designed to consume fissile or waste materials it may be higher. This and the lack of a moderator means that fast reactor cores are much smaller, with dimensions of the order of 1 m compared with 3 m for light-water reactors and 10 m for graphite or heavy-water reactors, and the power density is much higher.

In a fast reactor thermal neutrons are almost absent so the materials with high thermal neutron absorption cross-sections, which are so important in thermal reactors, do not affect the performance of a fast reactor nearly as much. Fission products such as ^{135}Xe and ^{147}Sm and impurities such as boron are relatively unimportant. There is no xenon poisoning problem for a fast reactor and the decrease of reactivity with burnup of the fuel due to the accumulation of fission products is much slower than in a thermal reactor. Because most materials have similar cross-sections for fast neutrons nuclear considerations place much less severe limits on the choice of materials for a fast reactor core.

The mean free path of fast neutrons is longer than that of thermal neutrons so the core of a fast reactor is more closely coupled than that of a thermal reactor. There is no question of zonal instability and there is less depression of the neutron flux in the fuel elements.

The temperature coefficients of reactivity come from entirely different sources – the Doppler effect and coolant expansion in fast reactors rather than moderator expansion and change in thermal energy in thermal reactors – but the magnitudes are similar so the dynamics of fast and thermal reactors are very similar in normal operation. Only in very rapid transients is there any difference because the prompt neutron lifetime is of the order of 10^{-7} s in a fast reactor, compared with about 10^{-3} s in a thermal reactor.

In spite of the simplicity of a fast reactor neutron flux calculations are much more complex because the simplifying assumptions valid for a thermal reactor cannot be made. In a thermal reactor most of the neutrons have energies in a narrow range and one-group or few-group calculations are useful. In a fast reactor the neutrons have a wide range of energies and multigroup calculations are essential. There is no fast reactor equivalent to the "four-factor formula".

1.1.3 Typical Reactors

The physics of a fast reactor depends on the materials that compose the core and on its size and shape but not strongly on the details of its structure, as explained in section 1.1.2. In broad terms the composition is determined as follows. The only material in the core that is essential to the physics is the fuel (unlike a thermal reactor in which the moderator is also essential), but the demands of heat transfer usually require that about 50% of the core volume is occupied by coolant. Structural material takes up another 20% or so leaving about 30% for fuel.

For a sodium-cooled power reactor in which the heat rating of the fuel is maximised the various demands of heat transfer and heat transport, which are discussed in Chapter 3, limit the height of the core to about 1 m and the average power density to about 500 MW m^{-3}. The diameter of the core is then determined by the required power output, and criticality is adjusted by changing the proportion of fissile material in the fuel. For gas-cooled or lead-cooled reactors the power density is lower, possibly about 200 MW m^{-3} or less, and the optimum core height is greater, possibly 1.5 m for gas coolant or 2 m for lead.

Most fast reactors that have been built so far, apart from experimental or test facilities, have been cooled with sodium and designed primarily to produce power.

1.2 CALCULATION METHODS

1.2.1 The Transport Equation

The neutron density n in a reactor is in general a function of position $\mathbf{r}$, the energy E of the neutrons, the direction $\mathbf{\Omega}$ in which they are travelling, and time t. The neutron density n obeys a linear version of the Boltzmann equation called the neutron transport equation, the derivation of which is given in many standard works on neutron transport (e.g. Davison and Sykes, 1957; Duderstadt and Hamilton, 1976). It can be written in a simplified form as

$$\frac{\partial n}{\partial t} = -v\mathbf{\Omega} \cdot \mathbf{grad}\, n - v\Sigma_r n + T + F + S.$$
$$\quad\quad\quad 1 \quad\quad\quad\quad 2 \quad\quad 3 \quad 4 \quad 5$$

$$(1.1)$$

If each term in equation 1.1 is multiplied by a small element dVdEd$\mathbf{\Omega}$ the terms on the right-hand side can be thought of as the contributions to neutrons appearing in the volume dV and the energy interval dE and travelling in a small solid angle d$\mathbf{\Omega}$ surrounding the direction $\mathbf{\Omega}$ as follows.

1. This is the rate at which neutrons already in dEd$\mathbf{\Omega}$ move into dV across its boundary, v being the neutron speed corresponding to E, so that $v^2 = 2E/m$ where m is the mass of a neutron. (Relativistic effects are not important: even a 4 MeV neutron is travelling at only a tenth of the speed of light.)
2. This is the rate at which the neutrons in dVdEd$\mathbf{\Omega}$ interact with the nuclei in dV. It is assumed that any interaction removes the neutron

from $dEd\mathbf{\Omega}$. Σ_r is the total macroscopic cross-section in $dVdE$ and is independent of $\mathbf{\Omega}$.

3. This is the rate at which neutrons already in dV are scattered (elastically or inelastically) from other energies and directions $dE'\,d\mathbf{\Omega}'$ into $dEd\mathbf{\Omega}$.

4. This is the rate at which new neutrons from fission appear in $dVdEd\mathbf{\Omega}$. Delayed neutrons are ignored in equation 1.1 for simplicity.

5. This is an additional source of neutrons in $dVdEd\mathbf{\Omega}$ (from spontaneous fission, for example, or from a spallation source driven by an accelerator).

T is given by

$$T = \int_{4\pi} d\mathbf{\Omega}' \int_0^\infty dE'\mathbf{v}'\Sigma_s(E' \to E, \mathbf{\Omega}' \to \mathbf{\Omega})n(\mathbf{r}, E', \mathbf{\Omega}', t). \quad (1.2)$$

Σ_s is the macroscopic scattering cross-section in dV (including both elastic and inelastic components). It is a function of E', E and $\mathbf{\Omega} - \mathbf{\Omega}'$. Up-scattering in energy is not important in a fast reactor so Σ_s is zero for $E' < E$.

F is given by

$$F = \int_0^\infty dE'n(E')\mathbf{v}'\Sigma_f(E')\bar{\nu}(E')\chi(E)/4\pi. \quad (1.3)$$

Σ_f is the macroscopic fission cross-section in dV and $\bar{\nu}$ is the average number of neutrons generated in each fission event. χ is the fission spectrum that is assumed to be independent of the energy of the neutron causing the fission. It is also assumed that fission neutrons are generated isotropically.

S is usually assumed to be zero in a critical reactor because in an operating power reactor spontaneous fission is negligible as a source of neutrons.

If $S = 0$ all the terms in equation 1.1 are linear in vn and it is usual to work in terms of the neutron flux ϕ defined by

$$\phi(\mathbf{r}, E, \mathbf{\Omega}, t) = \mathrm{v}(E)n(\mathbf{r}, E, \mathbf{\Omega}, t). \qquad (1.4)$$

v is about 14000 m s^{-1} at 1 eV and 14000 km s^{-1} at 1 MeV.

1.2.2 Discretisation

Equation 1.1 is far too complicated to be solved analytically without drastic simplification. Before it can be solved numerically it has to be recast with the continuous independent variables changed into discrete forms. This can be done in various ways, as follows. To make the explanation simpler we shall assume we are dealing with a reactor that is operating steadily so that we can ignore t.

Position. The position vector $\mathbf{r}$ is discretised in the form of a spatial mesh covering the reactor core and the surrounding breeder or consumer regions. It is usually convenient to make the geometry of the mesh coincide as far as possible with the actual structure of the core. As explained in Chapters 2 and 3, most fast reactor cores consist of hexagonal fuel subassemblies so a fully three-dimensional mesh is usually either hexagonal or triangular in the radial directions and linear in the axial. (This is in contrast to the square calculation meshes for thermal reactors, the cores of which are usually made up of square subassemblies.) For some purposes a cylindrically symmetric two-dimensional (r,z) approximation, linear in both directions, may be adequate.

Having defined the spatial mesh equation 1.1 can be recast in terms of the mesh intervals. There are two ways to do this. In the *finite difference* method ϕ is calculated at each mesh point, whereas in the *nodal* method the average value of ϕ over the volume of each cell of the mesh is calculated.

Energy. E is discretised by dividing the range of energy of interest (from about 10 MeV down to 0.1 eV or lower) into a series of intervals $E_0 - E_1$, $E_1 - E_2$, etc., where E_0 may be 10 MeV, and $E_1 < E_0$, $E_2 < E_1$, etc. The neutrons with energies between E_g and E_{g-1} are known as group-g neutrons. The flux ϕ_g in group g is then given by

$$\phi_g(\mathbf{r}, \mathbf{\Omega}) = \int_{E_g}^{E_{g+1}} \phi(\mathbf{r}, E, \mathbf{\Omega}) dE. \tag{1.5}$$

If there are G groups Σ_r, Σ_f and $\bar{\nu}$ are vectors with G elements. Σ_s becomes a G × G matrix the nature of which depends on how the directional variables $\mathbf{\Omega}$ are discretised.

Direction. The direction vector $\mathbf{\Omega}$ can be discretised in either of two ways. One is to specify a set of points on the surface of an imaginary unit sphere and then recast equation 1.1 in terms of the flux of neutrons moving in the directions of areas $d\mathbf{\Omega}$ surrounding these points. This leads to "S_n" methods, where n is the number of discrete directions. Alternatively ϕ can be expressed as the sum of an infinite series of spherical harmonic components, and equation 1.1 rewritten in terms of the coefficients of the series. In the case of a one-dimensional axially symmetric approximation

$$\phi = \sum_{n=0}^{\infty} a_n P_n(\cos\theta), \tag{1.6}$$

where the P_n are Legendre polynomials. This is called a "P_n" method. In either S_n or P_n calculations the greater the number of "mesh points" (i.e. the number of directions for S_n or terms of the series for P_n) the greater the accuracy. In practice $n = 5$ is usually found adequate.

1.2.3 The Diffusion Approximation

For the purposes of designing or operating a reactor the direction in which the neutrons are travelling is usually of little interest and there is an incentive to adopt a simpler calculation method that eliminates the two independent scalar variables $\boldsymbol{\Omega}$. This is the treatment known as "diffusion theory".

The basis of diffusion theory is to assume that ϕ is either isotropic or at most linearly anisotropic. (This is equivalent to P_1 transport theory, with all but the first two terms on the right-hand side of equation (1.6) being ignored.) The transport equation is integrated over all directions and various approximations are made to obtain an energy-dependent diffusion equation in terms of the neutron flux ϕ, independent of direction, defined by

$$\phi(\mathbf{r}, E) = \int_{4\pi} \phi(\mathbf{r}, E, \boldsymbol{\Omega})d\boldsymbol{\Omega}, \tag{1.7}$$

and the neutron current

$$\mathbf{J}(\mathbf{r}, E) = \int_{4\pi} \boldsymbol{\Omega}\phi(\mathbf{r}, E, \boldsymbol{\Omega})d\boldsymbol{\Omega}. \tag{1.8}$$

The assumption that the flux is at most linearly anisotropic is accurate except in a strongly absorbing medium or where the properties of the medium change substantially over distances comparable to the mean free path of the neutrons. For many fast reactor calculations these limitations are not particularly important. Fast neutron cross-sections are usually small and mean free paths are typically 0.1 m or longer. The nuclear properties of fuel, coolant and structure are very different but because the dimensions of individual fuel elements and structural members are usually only a few millimetres, over distances comparable with the mean free path large regions of the reactor can be treated as homogeneous and diffusion theory can be used. Even in control rods the capture cross-sections for fast neutrons are low enough for diffusion theory to be a good approximation for many purposes. The worst inaccuracies arise at the edge of the core, and deep in the neutron shielding that surrounds the reactor.

There are some types of reactors for which transport theory has to be used, such as the small experimental fast reactors of the 1950s and

1960s that had cores with dimensions comparable with the mean free path, and also experimental reactors assembled from various materials in the form of thin plates, typically 50 mm square, arranged in arrays comparable in size with the mean free paths.

The great advantage of diffusion theory over transport theory is that, because for a steady-state calculation it deals with four independent scalar variables rather than six, it makes smaller demands on computing power. For this reason up to the first decade of the 21st century most reactor design and operational flux calculations made use of diffusion theory. However in recent years the capacity of computers has increased to the extent that transport theory calculations have become more feasible and cheaper and therefore more widely used.

The rest of this discussion of fast reactor physics is presented in terms of diffusion theory because its formulation is much simpler than that of transport theory. Nevertheless it should be borne in mind that the same considerations apply in the latter, but the algebra is heavier.

1.2.4 Multigroup Diffusion Theory

After integration over all directions, some manipulation, and the discretisation of energy into G groups, the steady-state version of equation 1.1 for a homogeneous region of space gives G linked equations, one for each of the group fluxes ϕ_g,

$$0 = D_g \nabla^2 \phi_g - \Sigma_{rg}\phi_g + \sum_{g'=1}^{g-1} \Sigma_{sg' \to g}\phi_{g'} + \frac{1}{k}\chi_g \sum_{g'=1}^{G} \bar{\nu}_{g'} \Sigma_{fg'}\phi_{g'} + S_g. \quad (1.9)$$

Equations 1.9 are the multigroup diffusion equations. The terms on the right-hand side of equations 1.1 and 1.9 correspond to each other. The groups are numbered in reverse order of energy, so that group 1 contains the neutrons of highest energy and group G those of the lowest. Thus the limits of summation in the third term of equation 1.9 correspond to the absence of up-scattering. The diffusion coefficients D_g are given by $D_g = 1/3\Sigma_{sg}(1 - \bar{\mu}_g)$ where $\bar{\mu}$ is the mean cosine of the angle of scattering in the laboratory system of coordinates. The boundary conditions at an interface between regions with different properties are that ϕ_g and $\mathbf{J}_g = D_g \, \mathbf{grad} \, \phi_g$ should be continuous.

For a reactor driven by a source ($S_g \neq 0$) k is made equal to 1 and the ϕ_g represent the resulting flux distribution. If no solution is possible the reactor assembly is supercritical. For a critical reactor with $S_g = 0$ k can be thought of in two ways. The straightforward interpretation is that it is the effective multiplication constant, k_{eff}, of the reactor. But if $k \neq 1$ the reactor cannot be steady and the ϕ_g cannot be constant, so that under this interpretation equations 1.9 have a meaning for $k = 1$ only. Alternatively k can be thought of as the highest eigenvalue of the set of equations that can be solved to find k and the ϕ_g, although in this case the ϕ_g have no physical meaning unless $k = 1$. The composition of the reactor has to be altered to make $k = 1$, and when this is achieved the ϕ_g are the group fluxes in the critical reactor.

If the neutrons are treated in only one group – that is, if differences of energy can be ignored – the third term in equation 1.9 disappears and it becomes

$$0 = D\nabla^2\phi + (\overline{\nu}\Sigma_f/k - \Sigma_a)\varphi + S. \tag{1.10}$$

$\Sigma_a \equiv \Sigma_c + \Sigma_f$ has replaced Σ_r because neutrons are removed only by capture or fission.

Before the multigroup equations can be solved of course values have to be given to the constants D_g, Σ_{rg}, $\Sigma_{sg \to g'}$, $\chi_g \overline{\nu}_g$ and Σ_{fg} that depend on the microscopic cross-sections for the various materials in the reactor. Because the cross-sections vary with neutron energy the group constants involve average values over the energy range covered by each group.

In principle it is possible to make the groups so narrow that the variation of each cross-section within each group is small. However, in practice this cannot be done because the very complex fine structure of the variation of the cross-sections with energy, especially in the resonance region, would necessitate an impossibly large number of groups. A method has to be found for calculating suitable group cross-sections to facilitate accurate calculations with 20 or 30 groups.

1.2.5 Fundamental Mode Calculations

The difficulty of estimating how the flux varies within each group before the equations 1.9 are solved is overcome by making a very simple assumption about the spatial variation. When this assumption is made it is possible to estimate the detailed variation of the flux with energy (the "energy spectrum") and to use it to calculate group constants for a proper calculation of the spatial variation. In equations 1.9 the spatial variation is contained in the first term. The simplification is made by assuming that $\nabla^2 \phi_g$ can be replaced by $-B^2 \phi_g$ where B^2 is the "buckling", a constant that is the same for all energies.

The energy range is then divided into a large number of fine groups so that the variation of the cross-sections within each group is small. These fine groups can be denoted by the suffix n to distinguish them from the broad groups used in the spatial calculation, which are denoted by the suffix g. If B^2 is chosen suitably, corresponding to the system being critical, we can write $k = 1$.

Because we are dealing with only one point in space the fine group fluxes can be normalised so that

$$\sum_{n=1}^{N} v_n \Sigma_{fn} \phi_n = 1. \tag{1.11}$$

The equation for ϕ_n then becomes

$$\phi_n = \left\{ \sum_{n'=1}^{n-1} \Sigma_{sn' \to n} \phi_{n'} + \chi_n \right\} / (D_n B^2 + \Sigma_{rn}). \tag{1.12}$$

In this set of linear equations ϕ_n depends only on the $\phi_{n'}$ for $n' < n$ so they can be solved one after another starting with ϕ_1. It is then necessary to iterate, changing the value of B^2 until equation 1.11 is satisfied. This is the procedure of a "fundamental mode" or "normal mode" calculation. It can be carried out easily and quickly for a very large number of fine groups, typically 1000 or more, even though iteration is involved.

If the reactor consists of several regions with different compositions a separate fundamental mode calculation has to be done for each, obtaining different neutron spectra. For regions that are net absorbers of neutrons it is necessary to make $B^2 < 0$.

Once fine group spectra have been found for each region the broad group constants can be found. Thus, for example, the group fission cross-sections are given by

$$\bar{\nu}\Sigma_{fg} = \sum_{n\in g} \bar{\nu}_n \Sigma_{fn}\phi_n \Big/ \sum_{n\in g} \phi_n. \qquad (1.13)$$

Similarly the diffusion coefficient is given by

$$D_g = \sum_{n\in g} D_n \phi_n \Big/ \sum_{n\in g} \phi_n. \qquad (1.14)$$

This is based on the assumption that $\nabla^2\phi$ varies with energy in the same way as ϕ. This assumption is basic to a fundamental mode calculation.

The group transfer cross-sections include contributions from both inelastic and elastic scattering. The inelastic scattering component is given by

$$\Sigma_{isg'\to g} = \sum_{n\in g}\sum_{n'\in g'} \Sigma_{isn'\to n}\phi_{n'}/\phi_{g'}, \qquad (1.15)$$

but the elastic scattering has to be treated differently because if the broad groups are wide compared with ξE_g, where ξ is the mean logarithmic energy change, elastic scattering transfers neutrons from the lower end of the group only. Thus

$$\Sigma_{esg'\to g} = \sum_{n\in\ell(g-1)} \Sigma_{esn}\phi_n/\phi_{g-1}, \qquad (1.16)$$

where the summation is over the fine groups roughly within ξE_{g-1} of the group boundary.

As an alternative to doing a fundamental mode calculation it is possible to make a rough estimate of the group constants as follows. In a medium with no absorption and a constant scattering cross-section $\phi(E)$ would vary as $1/E$. Though the actual spectrum departs from this variation over wide energy ranges, over the width of a single group it is sometimes reasonably accurate to assume $\phi \propto 1/E$, giving

$$\bar{v}_g \Sigma_{fg} = \int_{E_g}^{E_{g-1}} \bar{v} \Sigma_f dE/EU_g \tag{1.17}$$

where

$$U_g = \ln(E_{g-1}/E_g). \tag{1.18}$$

Using this assumption the elastic scattering transfer cross-section is given by

$$\Sigma_{es(g-1)\to g} \simeq \xi \Sigma_{es} \phi(E_{g-1})/U_g. \tag{1.19}$$

1.2.6 Perturbation Theory

Although it is always possible to solve equations 1.9 to find the eigenvalue k of a system it is very useful to have a means of estimating the effect of small changes. This is particularly so in the case of temperature coefficients. For example, as explained later, a change in temperature makes small alterations to certain group cross-sections by means of the Doppler effect. A method of estimating the resultant change in k is needed, and this is provided by perturbation theory.

Suppose the perturbation in which we are interested results in an increase $\delta\Sigma_{cg}$ in the capture cross-section for group g in a small region dV at a point $\mathbf{r}$ in the reactor. The increased rate of capture of group g neutrons is then $\varphi_g(\mathbf{r})\delta\Sigma_{cg}\,dV$. The resulting effect on k depends on g and $\mathbf{r}$; that is to say neutrons in some parts of the reactor and at some energies are more important than others. For example neutrons captured at the edge of the core, which were anyway quite likely to have leaked out and been lost, are less important than neutrons captured at the centre, which were more likely to cause another fission. This distinction can be represented by assigning an importance, ϕ^*, to each neutron. ϕ^* depends on energy and position, so $\phi^* = \phi_g^*(\mathbf{r})$. ϕ^* can be defined in a simple physical way as follows (although it should be remembered that this is not the only way of interpreting it).

If a reactor with no source of neutrons apart from fission is exactly critical and operating at a certain mean power P, and a single neutron in group g is artificially introduced into it at a point $\mathbf{r}$, the power will, after a time, settle down to a new mean value $P + \delta P$, where δP is a random quantity. If it was possible to repeat the experiment many times the average change in power due to the introduction of one new neutron could, in principle, be found. This we call $\phi_g^*(\mathbf{r})$, the importance of the neutron in group g at $\mathbf{r}$.

With this definition in mind it can be seen that the rate of change of power due to the increased capture $\delta \Sigma_{cg}$ at $\mathbf{r}$ is

$$\dot{P} = -\phi_g^*(\mathbf{r})\phi_g(\mathbf{r})\delta \Sigma_{cg}\, dV. \tag{1.20}$$

To determine the effect on k we have also to know what the actual power is in terms of ϕ^*. The rate at which new group g neutrons are being generated in dV is $s_g\, dV$, where

$$s_g = \sum_{g'} \chi_g \bar{\nu}_{g'} \Sigma_{fg'} \phi_{g'}. \tag{1.21}$$

If the mean lifetime of these neutrons is λ the number of group g neutrons in dV is $\lambda s_g\, dV$ and their contribution to the reactor power is $\lambda \phi_g^* s_g\, dV$. The total reactor power is therefore

$$P = \sum_g \int_R \lambda \phi_g^* s_g\, dV = \int_R \sum_g \sum_{g'} \lambda \chi_g \bar{\nu}_{g'} \Sigma_{fg'} \phi_{g'} \phi_g^*\, dV \tag{1.22}$$

and the time constant is $\tau = P/\dot{P}$.

Elementary reactor kinetics using the point kinetics model shows that τ is related to reactivity ρ by $\tau = \lambda/\rho$, where ρ is defined as $(k-1)/k$. Thus we have finally

$$\rho = -\phi_g^*(\mathbf{r})\phi_g(\mathbf{r})\delta\Sigma_{cg}\,dV/C, \qquad (1.23)$$

where

$$C = \int_R \sum_g \sum_{g'} \chi_g \bar{\nu}_{g'} \Sigma_{fg'} \phi_{g'} \phi_g^* \, dV, \qquad (1.24)$$

the integral being taken to run over the whole of the reactor.

This treatment glosses over a lot of difficulties, chiefly that λ is different for neutrons in different groups, but it illustrates the principle. A rigorous treatment is given by Duderstadt and Hamilton (1976), for example, and shows that in the general case of perturbations to any of the group constants the change $\delta\rho$ in reactivity is given by

$$\delta\rho = \frac{1}{C}\int_R \left\{ \sum_g \sum_{g'} \left(\delta\left(\chi_g \bar{\nu}_{g'} \Sigma_{fg'}\right) + \delta\Sigma_{sg'\to g}\right)\phi_g^* \phi_{g'} \right.$$

$$\left. - \sum_g \left(\delta\Sigma_{rg}\phi_g^* \phi_g + \delta D_g \nabla\phi_g^* \cdot \nabla\phi_g\right) \right\} dV. \qquad (1.25)$$

In the case of a fundamental mode calculation the requirement that the reactor is exactly critical ($k = 1$) and that the total flux is normalised (equation 1.11) implies that the importance of neutrons born in fission is also normalised so that

$$\sum_{n=1}^{N} \chi_n \phi_n^* = 1. \qquad (1.26)$$

1.2.7 Matrix Notation

The complicated expression of equation 1.25, and indeed much of multigroup diffusion theory, can be written much more easily in matrix form. In this notation equations 1.9 become

$$\mathbf{M}\boldsymbol{\phi} = \frac{1}{k}\mathbf{F}\boldsymbol{\phi}, \tag{1.27}$$

where $\boldsymbol{\phi}$ is the column vector $\begin{pmatrix} \phi_1 \\ \phi_2 \\ .. \\ .. \end{pmatrix}$ and $\mathbf{M}$ and $\mathbf{F}$ are the matrices

$$\mathbf{M} \equiv \begin{pmatrix} -D_1\nabla^2 + \Sigma_{r1} & 0 & 0 & .. \\ -\Sigma_{s1\to 2} & -D_2\nabla^2 + \Sigma_{r2} & 0 & .. \\ -\Sigma_{s1\to 3} & -\Sigma_{s2\to 3} & -D_3\nabla^2 + \Sigma_{r3} & .. \\ .. & .. & .. & .. \end{pmatrix}, \tag{1.28}$$

and

$$\mathbf{F} \equiv \begin{pmatrix} \chi_1\bar{\nu}_1\Sigma_{f1} & \chi_1\bar{\nu}_2\Sigma_{f2} & \chi_1\bar{\nu}_3\Sigma_{f3} & .. \\ \chi_2\bar{\nu}_1\Sigma_{f1} & \chi_2\bar{\nu}_2\Sigma_{f2} & \chi_2\bar{\nu}_3\Sigma_{f3} & .. \\ \chi_3\bar{\nu}_1\Sigma_{f1} & \chi_3\bar{\nu}_2\Sigma_{f2} & \chi_3\bar{\nu}_3\Sigma_{f3} & .. \\ .. & .. & .. & .. \end{pmatrix}. \tag{1.29}$$

The neutron importance $\boldsymbol{\phi}^* = \begin{pmatrix} \phi_1^* \\ \phi_2^* \\ .. \\ .. \end{pmatrix}$ can be shown to be the solution of

$$\mathbf{M}^{\mathbf{T}}\boldsymbol{\phi}^* = \frac{1}{k}\mathbf{F}^{\mathbf{T}}\boldsymbol{\phi}^*, \tag{1.30}$$

where $\mathbf{M}^{\mathbf{T}}$ and $\mathbf{F}^{\mathbf{T}}$ are the transposes of $\mathbf{M}$ and $\mathbf{F}$ (i.e. the i, j element of $\mathbf{M}^{\mathbf{T}}$ is the j, i element of $\mathbf{M}$ etc.). $\mathbf{M}^{\mathbf{T}}$ and $\mathbf{F}^{\mathbf{T}}$ are actually the "adjoints" of $\mathbf{M}$ and $\mathbf{F}$, but because $\mathbf{M}$ and $\mathbf{F}$ are real the adjoint (which is the transpose of the complex conjugate) is the same as the transpose. $\boldsymbol{\phi}^*$ is often called the "adjoint flux", but this is misleading because it does not have the properties of a flux. In particular if two adjacent groups g and $g + 1$ are combined to form a new group g', $\phi_{g'} = \phi_g + \phi_{g+1}$, but in contrast ϕ^* is not additive and $\phi_{g'}^*$ is an average between ϕ_g^* and ϕ_{g+1}^*.

Equation 1.25, which is the basic result of first-order perturbation theory, in matrix notation is

$$\delta\rho = \int_R \boldsymbol{\phi}^*(\delta\mathbf{F} - \delta\mathbf{M})\boldsymbol{\phi}dv \bigg/ \int_R \boldsymbol{\phi}^*\mathbf{F}\boldsymbol{\phi}dv. \qquad (1.31)$$

This is called "first-order" for two reasons. The first is that second-order small quantities have been neglected in the usual way. The second is that $\boldsymbol{\phi}$ and $\boldsymbol{\phi}^*$ both refer to the unperturbed reactor. If in equation 1.31 $\boldsymbol{\phi}$ were the flux after the reactor had been perturbed the equation would be exactly right and the result would be that of exact perturbation theory. It would not be nearly so useful, however, because equation 1.31 makes it possible to estimate the effect of any small perturbation using the results of just two multigroup calculations, one to find $\boldsymbol{\phi}$ from equations 1.27, the other to find $\boldsymbol{\phi}^*$ from equations 1.30.

1.2.8 Resonances – the Effect of Temperature

The cross-sections of many nuclei exhibit sharp resonances over part of the energy range. Heavy nuclei such as uranium and plutonium have resonances in the range from about 1 eV up to 500 eV or more. Lighter elements have resonances mostly at higher energies of 100 keV or more, although there is an isolated resonance of ^{23}Na at 3 keV. It would be impossible to do a fundamental mode calculation for all these resonances even if they had been resolved experimentally because there would have to be far too many fine groups. Fortunately it is possible to calculate the reaction rate in a resonance directly, but in doing this it is important to allow for the effect of temperature. In outline the method is as follows.

In the vicinity of an isolated resonance the microscopic cross-section for a certain reaction, neutron capture say, is given by

$$\sigma_c = \sigma_{c0} \left(\frac{E_0}{E_c}\right)^{\frac{1}{2}} \bigg/ \left(1 + \frac{4(E_c - E_0)^2}{\Gamma^2}\right). \qquad (1.32)$$

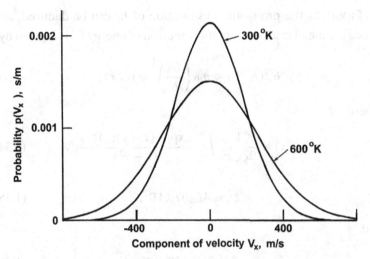

Figure 1.1 The effect of temperature on one component of the velocity of a ^{238}U nucleus.

Here σ_{c0} is the capture cross-section at E_0, the energy of the resonance peak, and Γ is the "width" of the resonance (the full width in energy at half the maximum cross-section). E_c is the kinetic energy of the neutron and nucleus relative to the centre of mass of the neutron-nucleus system. If the nucleus is stationary $E_c = E/(1 + 1/A)$ where A is the ratio of the mass of the nucleus to that of the neutron and E is the neutron energy.

Unless the reactor is at absolute zero temperature, however, the nucleus is unlikely to be stationary. It moves at random due to thermal agitation and this affects E_c, which is increased if the nucleus happens to be moving towards the approaching neutron or decreased if it is moving away. The effect on the apparent cross-section can be calculated if the probability distribution of the velocity of the nucleus is known. It is usually satisfactory to assume it to have a Maxwell-Boltzmann distribution. The resulting distribution of the component of velocity parallel to a certain direction (which we can take as the direction of the neutron) is shown in Figure 1.1 for an atom of ^{238} U at various temperatures.

From this the probability distribution of E_c can be deduced, and hence the mean cross-section for a neutron of energy E. It is given by

$$\bar{\sigma}_c(E, T) \approx \sigma_{c0} \left(\frac{E_0}{E}\right)^{\frac{1}{2}} \psi(\zeta, x), \tag{1.33}$$

where

$$\psi(\zeta, x) \equiv \frac{1}{2\zeta\sqrt{\pi}} \int_{-\infty}^{\infty} \frac{\exp(-(x-y)^2/4\zeta^2)}{(1+y^2)} dy, \tag{1.34}$$

$$\zeta^2 \equiv 4E_0 bT/A\Gamma^2, \tag{1.35}$$

and

$$x \equiv 2(E - E_0)/\Gamma. \tag{1.36}$$

T is the absolute temperature and b is Boltzmann's constant. The approximations used in reaching equation 1.33 introduce negligible errors in most fast reactor applications.

Figure 1.2 indicates how the effective cross-section depends on temperature. The parameters ψ, ζ and x are proportional to the cross-section, the absolute temperature and energy respectively.

1.2.9 Resonances – Effective Cross Sections

To determine the group cross-section for a neutron group containing a resonance we have to know how the flux varies in the resonance region. A simple approximation, which is fairly accurate in most cases, is the "narrow resonance approximation". Suppose we have a region made up of two materials, one of which has a constant scattering cross-section Σ_s while the other has only a single capture resonance. Suppose also that the resonance is narrow so that its width $\Gamma \ll \zeta E_0$, where ξ is the mean change in ln E in neutron scattering events. It can be shown that the flux $\phi(E)$ varies as $1/E\Sigma_t$ where Σ_t is the total cross-section – i.e. $\Sigma_t = \Sigma_s + N\sigma_c$ where N is the number of atoms of the capturing

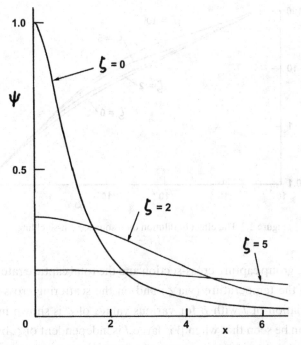

Figure 1.2 The effect of temperature on the effective cross-section in a resonance.

material per unit volume. The capture cross-section for a group that contains this resonance is then

$$\Sigma_{cg} = \int_{E_g}^{E_{g-1}} \frac{\overline{\sigma}_c dE}{\Sigma_t E} \bigg/ \int_{E}^{E_{g-1}} \frac{dE}{\Sigma_t E}, \tag{1.37}$$

and the denominator of this expression is ϕ_g, the total flux in the group. Substituting for Σ_t and making various approximations we obtain

$$\Sigma_{cg} \simeq \Gamma J/\phi_g, \tag{1.38}$$

where

$$J \equiv \frac{1}{2} \int_{-\infty}^{\infty} \frac{\psi \, dx}{\psi + \beta}, \tag{1.39}$$

and

$$\beta \equiv \Sigma_s/N\sigma_{c0}. \tag{1.40}$$

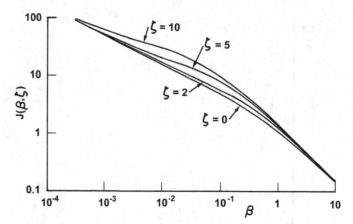

Figure 1.3 The effect of dilution on resonance self-shielding.

Thus the group capture cross-section and the total capture rate depend both on the temperature (via ζ) and on the scattering cross-section. The variation of J with β for various values of ζ is shown in Figure 1.3. It can be seen that when β is large J is independent of ζ, but when β is smaller J increases as ζ increases. This means that at "infinite dilution" when there is very little of the capturing material the capture rate is independent of temperature, but when more is present the capture rate increases as the temperature increases. This is known as the Doppler effect on the capture rate. It happens because as the temperature increases the effective resonance becomes lower and broader, as Figure 1.2 indicates. The flux at the peak of the resonance is less depressed, but this does not quite compensate for the lower cross-section so the reaction rate per unit energy ($\phi\sigma_c$) at the resonance peak decreases. At the sides of the resonance, however, σ_c increases more than ϕ decreases and $\phi\sigma_c$ increases. The increase at the sides outweighs the decrease at the peak and the total reaction rate ($\int \phi\sigma_c dE$) increases.

This applies to all resonance reactions, and of particular importance in a uranium-cycle fast reactor is the fact that both the fission rate in ^{235}U or ^{239}Pu and the capture rate, predominantly in ^{238}U, increase with temperature. The former tends to increase reactivity and the latter to decrease it. In a breeder reactor containing a large amount of ^{238}U the capture effect is greater and the resulting temperature coefficient of reactivity, known as the Doppler coefficient, is negative, making for stability. If the reactor is not designed to breed, however, and contains less ^{238}U or none at all this stabilising property is reduced and may be lost.

This discussion of resonance absorption and the Doppler effect is a simple version of the whole story. Other effects that have to be taken into account are the variation of the flux integral (the denominator in equation 1.37, which is not constant), the fact that resonances in one material are not in general isolated but overlap both with each other and with those of other materials, the fact that the narrow resonance approximation is not accurate, and the existence of unresolved resonances. The way these problems can be treated is described by Hummel and Okrent (1970).

1.2.10 Computation – Transport and Diffusion Theory

The information required to calculate the group fluxes and importances by solving equations 1.27 and 1.30 consists of the specification of the reactor and the nuclear cross-sections. The latter are available in the form of data from thousands of measurements stored in data libraries such as the USA Evaluated Nuclear Data File (ENDF), the OECD Joint Evaluated Fission File (JEFF), the Japanese Evaluated Nuclear Data Library (JENDL), the Chinese Evaluated Nuclear Data Library (CENDL) and the Russian Evaluated Nuclear Data Library (BROND). Microscopic cross-section data from these files are used, together with data about the specification, to calculate fine-group macroscopic cross-sections for each region. Fundamental-mode calculations in about 1000 fine groups are then performed to give group macroscopic cross-sections for spatial calculations, usually in 30 or 40 groups.

Equations 1.27 are solved by a double iterative method such as that described by Greenspan, Kelber and Okrent (1968). There are "inner" iterations to find the flux distribution and "outer" iterations to determine the eigenvalue, because the multigroup equations have no solution for ϕ until the correct value of k has been found. This outer iteration can be done in either of two ways. The composition and dimension of the reactor may be kept constant while the value of k is altered until the equations are solved. This is equivalent to finding the reactivity of a reactor that may not be exactly critical. Alternatively the composition (for example the concentration of plutonium in the core) or the dimensions (say the radius of the core) may be altered to make $k = 1$.

In the initial stages of design when the broad features of performance are being determined the three-dimensional reactor can often be represented adequately by a two-dimensional model in (r,z) cylindrical polar coordinates, but for the purposes of detailed design and for calculations in support of an operating reactor a three-dimensional model is required, usually in (hex,z) or (tri,z) geometry (i.e. with the transverse planes covered by a hexagonal or triangular mesh).

Again, in the initial stages of design, diffusion theory is adequate but when it comes to the details transport theory is necessary. A typical transport theory code would use a nodal formulation of the transport equation with hexagonal nodes, each node corresponding to an individual core position occupied by a fuel subassembly, a control rod or an incineration target, etc. The code would calculate the average flux in each node and then determine the fine structure within the node in terms of spherical harmonics in angle and polynomials in space. The average flux would enable properties such as the power generated in the subassembly to be predicted, and the fine structure would give the power of individual fuel elements within the subassembly.

A typical approach to operational calculations is given by Wardleworth and Wheeler (1974).

1.2.11 Computation – the Monte Carlo Method

Transport theory and diffusion theory are based on equations that describe the average behaviour of large numbers of neutrons as they interact with large numbers of nuclei. As the preceding paragraphs show the equations are complex and their solution by numerical means even more so. This is in contrast to the actual behaviour of individual neutrons, which is quite simple: they travel in straight lines between nuclei and they interact with the nuclei in a limited number of ways the probabilities of which are known. The transport equation has the effect of turning a large number of simple problems into a single complex problem. The alternative, which is to solve the simple problems, is the "Monte Carlo" method (Brown, 2012).

The basis of the method is to track neutrons one by one. The outcome of each event in the history of each neutron is chosen at random in accordance with the known probabilities. This is done for large numbers of different neutron histories, so large a number that when they are all put together they make up an estimate of the actual neutron distribution.

The method depends on a random number generator that produces a sequence of numbers R_n distributed uniformly at random in the range (0,1). For a "source-type" calculation (for example a subcritical reactor driven by a neutron source) the procedure is straightforward. A neutron is assumed to emerge from the source travelling in a direction (θ, ψ) chosen at random (which is ensured by setting $\psi = 4\pi R_1$ and $\theta = \cos^{-1}(1 - 2R_2)$) and with an energy E given by $R_3 = \int_0^E s(E')dE'$, s being the spectrum of source neutrons normalised to 1. The reactor is assumed to be made up of discrete regions (fuel, coolant, structure, etc.) each of which has a uniform composition and uniform nuclear properties. The neutron is assumed to travel in its initial direction for a number $m = -\ln R_4$ of mean free paths to its first interaction with a nucleus. In its course it may well cross one or more boundaries from region to region. The actual distance travelled is given by $m = x_1 \Sigma_{t1} + x_2 \Sigma_{t2} + \ldots x_i \Sigma_i$ where x_1 is the distance travelled in the region in which it was born, x_2 the distance through the next region, and so on, and x_i is the distance travelled in the region i in which it finally interacts with a nucleus.

What happens when it interacts is then decided by the use of further random numbers to select the type of nuclide it interacts with and the interaction (elastic scattering, inelastic scattering, capture or fission) that takes place, the choices being weighted by the various macroscopic cross-sections. In the case of elastic or inelastic scattering the neutron continues on its way in a different direction and with a different energy, again chosen at random taking account of any possible anisotropy of elastic scattering and the probabilities of exciting different nuclear energy levels in the case of inelastic scattering. In the case of capture the neutron disappears. In the case of fission it is replaced by one or more new neutrons, their number and energies chosen at random in accordance with data on $\bar{\nu}$ and χ for the fissile nuclide concerned. These new neutrons are then tracked as before and the tracking continues until all daughter neutrons are captured or have travelled out of the reactor.

The same procedure is followed for many neutrons and accounts are kept of the numbers and locations of the different interactions. For example reaction rates can be estimated by keeping account of the number of reactions taking place in an element of volume, and the power density can be deduced from the rate at which fissions take place. Neutron flux can be estimated by keeping account of the total path-length traversed by neutrons in the volume element, and neutron current by noting the number of neutrons that pass through an element of area. An estimate of one of these quantities is given by $\lim_{N \to \infty} (N_q/N)$, where N_q is the number of records of event q and N is the total number of neutron histories that have been calculated. Additional neutrons are tracked until a statistical test shows that the variance of the estimates of q is small enough to give confidence that

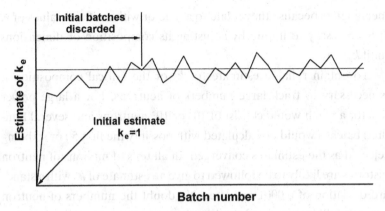

Figure 1.4 Monte Carlo calculations: estimating k_e.

the limit has been approached sufficiently closely. The most intricate part of the calculation is the determination of which regions a neutron passes through and where it crosses the inter-region boundaries.

The procedure for a critical reactor is a little more complicated than that for one driven by a neutron source because as well as the fluxes and reaction rates the eigenvalue k_e has to be estimated. It can be done as follows. A number N_1 of neutrons, sometimes called a "batch", are tracked up to the point at which they are captured, leak out of the reactor, or cause fissions. If a total of N_2 new neutrons are born in these fissions, $N_2/N_1 = k_{e1}$ is a first estimate of k_e. All values of v are then divided by this estimate and the N_2 new neutrons from the fissions caused by the first batch, which become the second batch, are tracked until they in turn cause fissions that give rise to N_3 further neutrons. $(N_3/N_2)k_{e1} = k_{e2}$ is then a new estimate of k_e, the v are modified again, a third batch of neutrons are tracked, and so on. In due course the estimates cluster around a limit, which is the required value of k_e, and the process continues until the variance is small enough. In making the final estimate of k_e it is normal to reject the values produced by the initial batches of neutrons as they converge towards the final value. The process is shown diagrammatically in Figure 1.4. The neutron fluxes generated in the course of this process are of course meaningless because they relate to a reactor with artificial values of ν. It is necessary to iterate by adjusting its composition or dimensions until $k_e = 1$.

To obtain reliable estimates of k_e or the critical composition it is necessary to track large numbers of neutrons. For a large power reactor a batch would consist of 10^4 or 10^5 neutrons and several hundred batches would be calculated with possibly the first 50 or so being rejected as the estimates converged. In all tens of millions of neutron histories are likely to be followed to give an estimate of k_e with a standard deviation of 0.0001 or better. No doubt the numbers of neutron histories will increase and the k_e estimates will improve as computing facilities become more powerful. (The numbers can be put in perspective by remembering that a 2500 MW (heat) reactor is producing some 2×10^{20} fission neutrons per second.)

The variance of the estimate of a quantity such as the flux at a certain point in the reactor depends on the number of neutron histories that contribute to it. This implies that more histories have to be calculated for a reliable estimate of the flux in a low-flux region than for the flux at the centre of the core. Since the variance is roughly inversely proportional to the square root of the number of histories this implies that if it is necessary to calculate N histories to attain a satisfactory estimate of the flux at the core centre, $100\,N$ histories will be required for an equally reliable estimate of the flux in a region, such as a breeder or a reflector, where it is a tenth of that at the centre. This consideration would, in principle, make it impossible to use a Monte Carlo method to calculate the performance of a reactor shield, which may be required to attenuate the flux by a factor of 10^{12} or more, but the difficulty can be avoided by "splitting" neutrons.

For example if a neutron crosses the outer boundary of the reactor core and enters the shield it can be replaced by two neutrons each of which has half the "weight" of the original neutron and is then allowed to have an independent history. The process can be repeated at other boundaries in the outer parts of the shield so that there are statistically

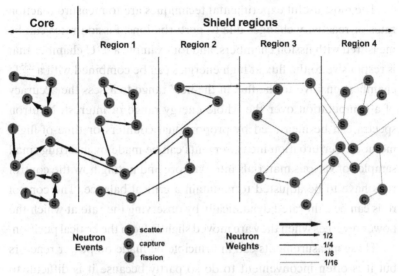

Figure 1.5 Monte Carlo calculations: "splitting" neutrons.

significant numbers of neutrons throughout the shield but the attenuation is accounted for by their diminished weights. The process is shown diagrammatically in Figure 1.5.

1.2.12 Accuracy and Experimental Checks

Multigroup diffusion calculations can normally be expected to predict the reactivity of a system within about 0.5% and the distribution of power within about 4% throughout most of the core and within 6% close to control rods or the edge of the core. The errors are due partly to uncertainties in the basic cross-section data and partly to the approximations inherent in diffusion theory. The latter explain the larger errors near interfaces between different regions.

It is possible to check calculations by means of measurements in a reactor. In the past such experimental checks were very valuable but they have become less important as the reliability of calculation methods has been confirmed.

The most useful experimental techniques are to measure reaction rates or reactivity changes due to perturbations. Fission rates can be measured with fission chambers, and for example a ^{238}U chamber that is responsive to the flux at high energies can be combined with a ^{235}U chamber sensitive to the flux in the keV range to assess the accuracy of a computation over the whole energy range of interest. Neutron spectra can be measured by proportional counters or time-of-flight methods. Perturbation measurements can be made by inserting small samples of various materials into the core and noting how the control rods have to be adjusted to maintain a critical balance. The control rods can be calibrated dynamically by observing the rate at which the power diverges when they are moved slightly from the critical position.

These measurements can in principle be made on power reactors but it is often inconvenient to do so partly because it is difficult to obtain access to the core, partly because the temperature and neutron flux in the core may be too high for the measuring instruments, and partly because performing reactor physics experiments conflicts with use of the reactor for the purpose for which it was built. It is very difficult to make a perturbation measurement at anything other than very low power, for example.

For this reason most, but not all, experimental checks on reactor physics calculations have come from zero power reactors. The most productive have been demountable facilities in which a reactor of almost any required composition could be assembled from samples of the various materials – structure, fuel, and even sodium coolant – present in a power reactor. Test reactors of this type operated at a power of typically a few watts so that no cooling was needed. The use of zero-power experimental reactors is described in more detail by Broomfield et al. (1969).

Figure 1.6 shows calculated and measured neutron energy spectra for a zero-energy experimental assembly (Broomfield et al., 1969). The assembly was similar in composition and size to a power reactor except that it contained carbon instead of oxygen and sodium. The spectrum

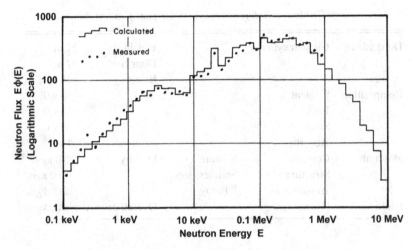

Figure 1.6 Experimental and calculated neutron energy spectra.

was measured by a time-of-flight method, and is compared with the
result of a 46-group fundamental mode calculation. The depression in
the flux caused by the iron resonance at 30 keV can be seen clearly.
If there had been sodium in the assembly there would have been
another depression due to the important sodium resonance at 3 keV.
The quantity plotted on a logarithmic scale in Figure 1.6 is the flux per
unit lethargy, $\phi(U)$. "Lethargy" U is defined by $U = -\log E$, where E
is the neutron energy. If $\phi(E)dE$ is the flux of neutrons with energy in
the range $E \rightarrow E + dE$ then $\phi(U) = 2.303E\phi(E)$.

1.3 NEUTRON FLUX

1.3.1 Energy Spectra

To illustrate the effects of various design options on the flux and
importance spectra, in the following paragraphs comparisons are made
with the "reference core" described in Table 1.1, which is a simplified
representation of a 2500 MW (heat) breeder reactor. The "absorber"
component represents the effect of the control rods when the fuel is

Table 1.1 *Reference reactor core specification*

Dimensions	Circular cylinder		Height	1 m
			Diameter	2 m
			Buckling	18 m^{-2}
Composition	Coolant			50 v/0
	Fuel			30 v/0
	Structure			19 v/0
	Absorber			1 v/0
Materials	Coolant	Sodium	Density	840 kgm^{-3}
	Structure	Stainless steel	"	7900 kgm^{-3}
	Absorber	^{10}BeO$_2$	"	2000 kgm^{-3}
Fuel	(U,Pu)O$_2$		Overall density	8900 kgm^{-3}

new or of the accumulated fission products at the end of its irradiation life. The stainless steel structure is assumed to be 74% Fe, 18% Cr and 8% Ni. The overall density of the fuel material is 80% of the theoretical density of the oxide, allowing for porosity incorporated to accommodate fission products (as explained in Chapter 2). The spectra are derived from a fundamental mode calculation based on the ANL 16-group cross-section data (Tamplin, 1963). The enrichment E, defined as Pu/(U+Pu), is adjusted to make the reactor critical.

Figure 1.7 shows the spectrum of the flux in the reference core compared with the spectrum of neutrons as they are born in fission of ^{239}Pu. (The fission spectrum for other isotopes is very similar.) The energy of the fission neutrons is reduced by scattering so that the peak is at around 0.3 MeV, and below this energy the flux falls off steeply until there are hardly any neutrons with energies less than about 1 keV. The spectrum in a fast reactor is very different from that in a thermal reactor.

At high energies, above about 0.5 MeV, inelastic scattering in ^{238}U and to a lesser extent in ^{56}Fe and ^{23}Na is very important. Excitation of the lowest energy level of the ^{238}U nucleus reduces the energy of a neutron by 45 keV and the corresponding values for ^{56}Fe and ^{23}Na are 845 and 439 keV respectively, so the effect of inelastic scattering is

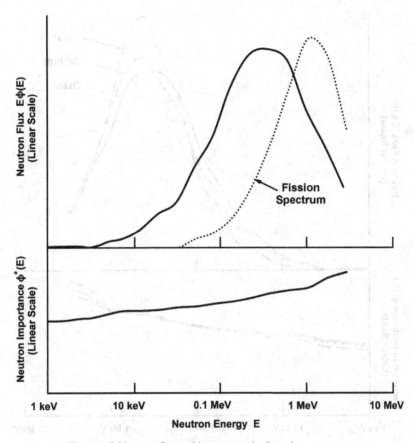

Figure 1.7 Neutron flux and importance in the reference core.

very marked. At lower energies it is elastic scattering that reduces the energy of the neutrons and the lightest nuclei such as ^{23}Na, ^{16}O and ^{12}C are the most important moderators. Many collisions are needed, however, to reduce the energy, and the chance that the neutron will diffuse out of the core or be absorbed is large, so the flux declines steadily with decreasing energy. A few of the neutrons are captured in ^{23}Na or ^{56}Fe but most are absorbed in ^{238}U.

The neutron importance, also shown in Figure 1.7, increases with energy. Above 1 MeV it is high because of the possibility of fission in ^{238}U. At low energies, below 3 keV, (not shown in Figure 1.7) it

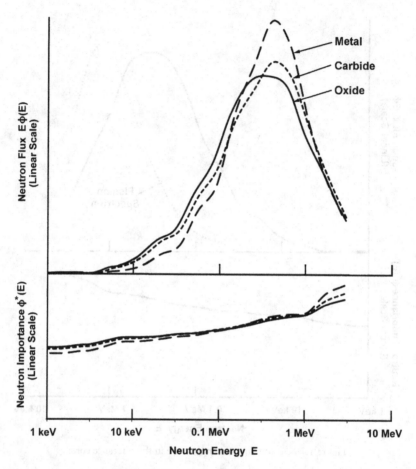

Figure 1.8 The effect of the fuel material.

rises with decreasing energy because the fission cross-section in ^{239}Pu rises more rapidly than the capture cross-section in ^{238}U, so the lower its energy the more likely a neutron is to cause fission and therefore contribute to the reactivity. It should be remembered that a neutron captured in ^{238}U is lost as far as maintaining the chain reaction is concerned even though it causes the generation of a new fissile nucleus.

Fuel Material. The effect of replacing oxide fuel in the reference core by carbide or metal is shown by comparing the spectra in Figure 1.8. (The overall densities are 10900 and 14300 kg m^{-3} respectively, both 80% of theoretical.) Fewer moderating atoms are present in (U,Pu)C than in (U,Pu)O_2 and even though a carbon nucleus is lighter than oxygen there is less moderation so the mean neutron energy is higher, as indicated by the fact that the peak in the spectrum is at a higher energy. This is usually called a "harder" spectrum. At the peak of the spectrum neutron importance increases with energy so the harder spectrum resulting from a change from oxide to carbide fuel allows fissile material to be replaced by fertile. As a result of this, and the higher fuel density, the enrichment is lower and there are more captures in the fertile material in the core. The importance is higher above 1 MeV because the enrichment is lower and more ^{238}U is present, and it is lower at lower energies because the ratio of fissile material to absorbers is lower.

Similar, but greater, changes result if oxide is replaced by metal as the fuel material.

Coolant. The choice of coolant has a great effect on the neutron flux and the performance of the reactor mainly because it occupies such a large fraction of the core volume. The effect is mainly due to differences in the inelastic scattering at energies above about 1 MeV. Figure 1.9 shows the spectra for reference cores cooled by sodium, lead-bismuth eutectic (54.5%Pb, 45.5%Bi) and carbon dioxide. Lead-bismuth eutectic and carbon dioxide are much poorer moderators than sodium (lead-bismuth because the atoms are heavier, carbon dioxide because it is much less dense), so in both cases the flux spectrum is harder. The fission-neutron spectrum peaks at around 2 MeV (see Figure 1.7). Strong inelastic scattering in ^{238}U and iron scatters many of these neutrons down to around 0.7 Mev, causing the sharp peaks in the lead-bismuth and carbon dioxide spectra. However elastic scattering in sodium is particularly strong at this energy so this peak is "smoothed out" in the sodium spectrum.

Although lead-bismuth is a poor moderator it has a high macroscopic scattering cross-section, whereas that of a gas coolant is very low. As a result the former leads to a lower critical enrichment than

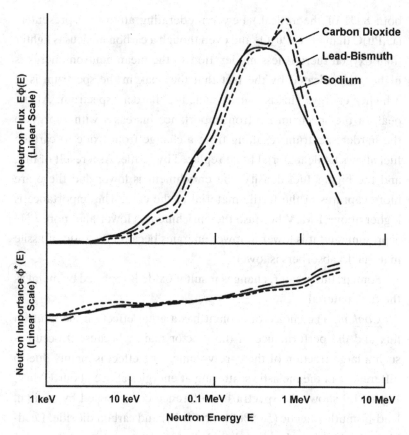

Figure 1.9 The effect of the coolant.

sodium and the latter to higher. Largely as a result of this the neutron importance curve is steeper for lead-bismuth and shallower for carbon dioxide.

Core Size. Figure 1.10 shows what happens if the core of the reactor is made smaller. It compares the spectrum for reference cores with $-B^2 = 18 \text{ m}^{-2}$ (a cylinder 1 m high and 2 m in diameter) and $-B^2 = 28 \text{ m}^{-2}$ (0.9 m high and 1.2 m in diameter). The difference is that 47% of the fission neutrons leak from the smaller core whereas only 32% leak from the larger. As a result the spectrum is harder and, as in the

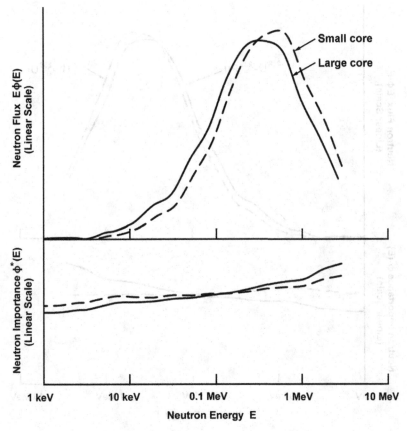

Figure 1.10 The effect of core size.

case of a gas-cooled core, the importance curve is shallower because the critical enrichment is higher.

Plutonium Composition. These comparisons have been made assuming the fissile material in the core is pure ^{239}Pu, but this is in practice very unrealistic. The plutonium is most likely to originate from the reprocessing of irradiated thermal reactor fuel. In the thermal reactor higher plutonium isotopes are formed by successive neutron capture reactions, and the longer it is irradiated – i.e. the higher the burnup – the more of them there are. For example plutonium from AGR fuel irradiated to 20000 MWd per tonne consists of Pu-239, 240, 241 and

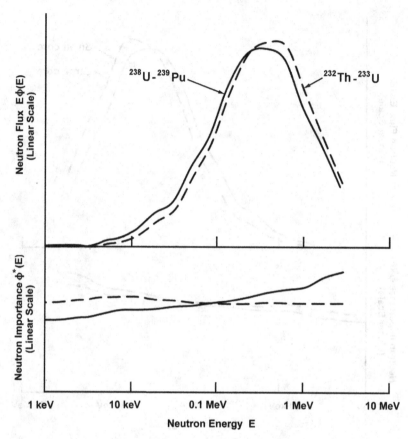

Figure 1.11 The effect of the thorium cycle.

242 in the proportions 56:20:15:9. If plutonium of this isotopic composition is substituted for pure ^{239}Pu in the reference reactor there is very little change in the spectrum but there is a large effect on the enrichment, which goes from 26% to 30%. However this is rather misleading because the enrichment is defined as (total Pu)/(Pu+U), and "total Pu" now contains a significant amount of fertile material. The actual ratio fissile/(fissile + fertile) falls to 22%. This is because ^{241}Pu has a higher fission cross-section than ^{239}Pu and a higher value of $\bar{v}$.

Thorium. Figure 1.11 compares reactors utilising the uranium and thorium cycles by showing the effect of replacing ^{239}Pu and ^{238}U with

Table 1.2 *Neutron balances*

	Ref. Core	Metal Fuel	Carbide Fuel	Pb/Bi Coolant	Gas Coolant	Small Core	AGR Plut.	Thorium Cycle
Enrichment (%)	25.8	14.6	20.1	22.2	28.0	34.7	30.8	26.8
Neutron Production								
Fissile Nuclides	0.904	0.836	0.875	0.913	0.902	0.922	0.813	0.982
Fertile Nuclides	0.096	0.164	0.125	0.087	0.098	0.078	0.187	0.018
Neutron Consumption								
Absorption in Fuel	0.488	0.538	0.514	0.513	0.462	0.444	0.486	0.519
Capture in Coolant	0.001	0.001	0.001	0.005	0.000	0.000	0.001	0.001
Capture in Structure	0.012	0.010	0.011	0.014	0.010	0.009	0.012	0.011
Capture in Absorbers	0.125	0.093	0.115	0.147	0.098	0.083	0.126	0.106
Leakage	0.374	0.358	0.359	0.321	0.430	0.464	0.375	0.363
Breeding Ratio	1.08	1.34	1.17	1.02	1.19	1.19	1.26	0.94

^{233}U and ^{232}Th. The flux spectrum is harder in the thorium case because there is less inelastic scattering at high energy. The enrichment is higher (27% compared with 23%) because the fission cross-section of ^{233}U is lower than that of ^{239}Pu in the 0.1–1.0 MeV range where the flux peaks. However there is a very significant difference in the importance spectra, which for the thorium reactor is almost completely flat. This is because the fission cross-section of ^{233}U is significantly higher than that of ^{239}Pu at lower energies.

Neutron Balance. Table 1.2 shows the sources and fates of the neutrons in these cores and illustrates the effects of the different spectra. As compared with a thermal reactor, fast fission in ^{238}U, and in the cases where they are present, ^{232}Th, ^{240}Pu and ^{242}Pu, is a more significant source of neutrons, and loss of neutrons by capture in the coolant or the structure is quite insignificant. In contrast many neutrons leak from

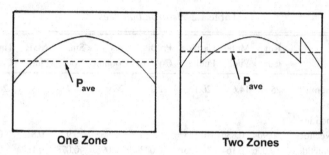

One Zone   Two Zones

Figure 1.12 The effect of enrichment zones on the radial distribution of power.

the core, but these are not necessarily lost because they can be made use of if the core is surrounded either by a breeder containing fertile material or by waste materials to be eliminated by transmutation.

An oxide-fuelled thorium-cycle reactor with this specification would not breed because the capture cross-section of ^{232}Th is lower than that of ^{238}U in the keV energy range, but a higher breeding ratio can be attained with carbide fuel.

1.3.2 Power Distribution and Enrichment Zones

If the core had the same composition throughout the power density would be nearly proportional to the flux and would be distributed across the core as shown on the left of Figure 1.12. This would be most undesirable from a thermodynamic point of view because if the coolant flowed at the same rate through all parts of the core it would emerge at different temperatures. When cold coolant from the periphery of the core mixed with hot coolant from the centre there would be a gain of entropy and a consequent loss of work output, and there would be large temperature fluctuations in the mixing region that might damage the structure (see Chapter 3). Alternatively if the flow in the outer part of the core were restricted to equalise the outlet temperatures pumping work would be wasted as the coolant flowed through the restrictions. For this reason the core is normally made in two or more radial zones, with higher fuel enrichment in the outer zones. The effect of this is to increase the power density in the outer region as shown on

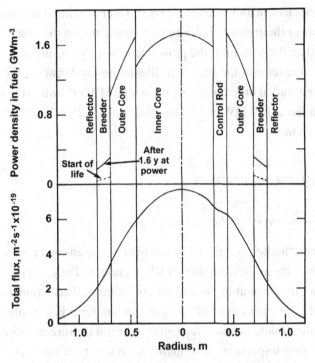

Figure 1.13 The radial distribution of flux and power density.

the right of Figure 1.12, thus reducing the coolant outlet temperature differences.

Figure 1.13 shows the radial distribution of neutron flux and power density in a small breeder reactor having two core zones, of roughly equal volume, with enrichments of 22% and 28%, surrounded by a breeder. If the core were uniform the enrichment would be about 24%. The peak power densities at the centre of the core and the inside of the outer zone are roughly the same and the radial power peaking factor, P_{max}/P_{ave}, where P_{max} is the power generated in the most highly rated channel and P_{ave} is the average power per channel, is reduced from about 1.35 in a single-zone core to 1.21 in the two-zone core.

An important point to notice in Figure 1.13 is the change in the power density at the inside of the breeder with time. As fissile material is generated in the breeder, predominantly at its inside edge, it in turn undergoes fission and generates power. Over the life of the inner fuel elements of the radial breeder (which would stay in the reactor much longer than the core fuel) the power density rises considerably. In this particular reactor at the start of its life the power density at the centre of an inner radial breeder fuel element is 60 MW m^{-3} when the reactor is operating at 600 MW (heat), but this rises to 220 MW m^{-3} after 1.6 years at power.

1.4 HIGHER ACTINIDES

1.4.1 Formation of Higher Actinides

The term "higher actinides" is used for the man-made nuclides of elements with atomic number of 93 or greater. They originate from neutron capture in uranium and are sometimes called "trans-uranium elements" or "trans-uranics". In practice the term is also often used for the man-made nuclides formed by neutron capture in thorium.

The most important of the higher actinides, both for reactor operation and commercially, are ^{239}Pu and ^{233}U produced by neutron capture in ^{238}U and ^{232}Th respectively, as explained in the Introduction. They are both beginnings of long and complicated chains of reactions, mainly neutron captures and β decays, that produce many isotopes of several elements. Figures 1.14 and 1.15 are simplified diagrams of these chains. The horizontal arrows represent neutron captures and the vertical ones β$^-$ (upward) or β$^+$ (downward) decays.

Figures 1.14 and 1.15 include the reactions that are of greatest practical importance, but many details have been omitted in the interest of clarity. For example ^{243}Am is β-active but this is not significant because its half-life is 7370 years. Almost all the nuclides in both diagrams that are not β-active are subject to α-decay with half-lives of greater than 1000 years so they are of little importance as far as reactor operation is concerned. (They are much more important in reprocessing and waste storage because the radiation is hazardous.) ^{241}Am however has

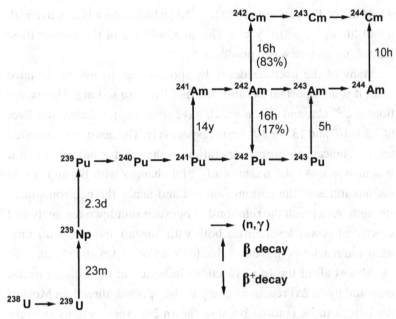

Figure 1.14 The formation of higher actinides from ^{238}U.

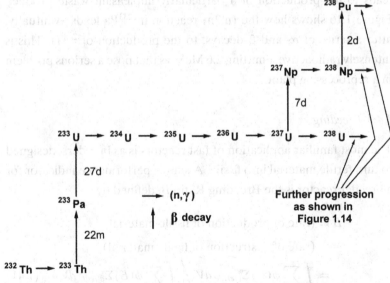

Figure 1.15 The formation of higher actinides from ^{232}Th.

a half-life of 433 years to produce ^{237}Np, which in turn is α-active with a half-life of 2.1×10^6 years. This makes it one of the longest lived hazardous nuclear waste products.

Many of the nuclides decay by spontaneous fission but in most cases it is unimportant because the half-lives are so long. The exceptions are ^{242}Cm and ^{244}Cm which have spontaneous fission half-lives of 7.2×10^6 and 13.2×10^6 years respectively. The neutrons generated are insignificant in normal operation of the reactor but not when it is shut down. As the quantity of ^{241}Pu changes with burnup, so do the quantities of the curium isotopes and hence the neutron source strength. As a result the relationship between shutdown reactivity and subcritical power level varies both with burnup and also with time while the reactor is shut down due to decay of ^{242}Am and ^{244}Am.

Almost all of the (n,γ) reactions indicated in both diagrams are mirrored by (n,2n) reactions going in the opposite direction. Most of the latter can be ignored because the (n,2n) cross-sections are very small. However, there is one that is of some significance because it leads to the production of a particularly unpleasant waste product. Figure 1.16 shows how the (n,2n) reaction in ^{233}Pa leads eventually, after a series of α- and β-decays, to the production of ^{208}Tl. This is intensely radioactive, emitting 2.6 MeV γs that pose a serious problem in a reprocessing plant.

1.4.2 Breeding

The most familiar application of fast reactors is as breeders, designed to turn fertile material into fissile. A simple performance indicator for a breeder reactor is the Breeding Ratio B, defined by

$$B = \text{(rate of production of fissile material)}/$$
$$\text{(rate of destruction of fissile material)}$$
$$= \int \sum \phi(E) \Sigma_{cfertile} \, dV \bigg/ \int \sum \phi(E) \Sigma_{afissile} \, dV \qquad (1.41)$$

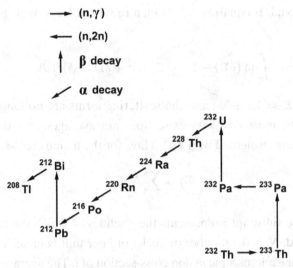

Figure 1.16 The formation of ^{208}Tl from ^{232}Th.

where $\Sigma_{cfertile}$ is the macroscopic cross-section for neutron capture in fertile material and $\Sigma_{afissile}$ for absorption in fissile. The sums run over all energy groups and the integrals over the entire reactor including the breeder region.

B is widely used as a measure of the effectiveness of a breeder reactor, but it suffers from the major disadvantage that it does not take account of the differences between the isotopes. It counts an atom of ^{241}Pu as being equivalent to an atom of ^{239}Pu, whereas in fact the fission cross-section of ^{241}Pu is higher and its capture cross-section is lower, so it is more valuable as a reactor fuel. A better measure of breeding is the "Breeding Gain" introduced by Baker and Ross (1963). This is based on an assessment of the values of the various isotopes as contributors to reactivity using perturbation theory in a simplified form involving a single energy group.

If there is only one neutron energy group we can see from equations 1.27 and 1.30 that neutron flux and importance, ϕ and ϕ^*, are proportional. If equation 1.25 is then rewritten for a single group we have

$$\delta\rho \propto \int_R \left(\delta\left(\bar{\nu}\Sigma_f - \Sigma_f - \Sigma_c\right)\phi^2 + \delta D\left(\nabla\phi\right)^2\right) dV \qquad (1.42)$$

because $\Sigma_r = \Sigma_f + \Sigma_c$, and the scattering terms are no longer significant. The macroscopic cross-sections are averages over the whole energy range weighted with $\phi\phi^*$. Thus, for the fission cross-section,

$$\Sigma_f = \sum_i N_i\bar{\sigma}_{fi}, \qquad (1.43)$$

where the subscript i represents the nuclides of which the reactor is composed, N_i is the number of nuclei of i per unit volume, and $\bar{\sigma}_{fi}$ is the average microscopic fission cross-section of i. The average is taken over the entire neutron energy range, weighted with $\phi\phi^*$; i.e.

$$\bar{\sigma}_{fi} = \int_0^\infty \sigma_{fi}(E)\phi(E)\phi^*(E)dE \Big/ \int_0^\infty \phi(E)\phi^*(E)dE, \quad (1.44)$$

where $\phi(E)$ and $\phi^*(E)$ are the fundamental mode flux and importance.

It follows from equation 1.42 that, if the effect of scattering is neglected, the reactivity increase when an atom of i is created is proportional to w_i, where

$$w_i = \overline{\nu\sigma}_{fi} - \bar{\sigma}_{fi} - \bar{\sigma}_{ci}. \qquad (1.45)$$

w_i is the "worth" of an atom of i and measures its usefulness for building a new reactor of the same design. The "Breeding Gain", G, is then defined as the net increase in worth (summed over all the nuclides present in the reactor) divided by the worth of the nuclides destroyed. It is still rather artificial because it assumes that the changes are distributed throughout the reactor in proportion to the numbers of nuclides present. (Thus fissile nuclides generated outside the core in a breeder region are assigned the worth they would have if they were in the core.) It nevertheless indicates the rate at which new cores, of the same specification, could be assembled when all the core and breeder fuel is reprocessed.

It is clear from the way they are defined that $G \approx B - 1$, but there is no algebraic way to show this to be the case.

1.4.3 Internal Breeding

G measures the gain in reactivity assuming the fuel is to be used in a new reactor. Breeding also has an effect on the reactivity of the reactor in which it takes place, but this is not related at all closely to the value of G because much of the new fissile material is generated in the breeder where its effect on reactivity is very small. The fissile material generated within the core does have a significant effect, however, which is very important. It can be characterised by the "internal breeding gain", G_I, which is defined in exactly the same way as G except that the integral in equation 1.42 runs over the core of the reactor alone, not over the whole reactor including the breeder. G_I is negative for most reactors, implying that there is a net loss of fissile material from the core, but it can be positive for metal-fuelled or gas-cooled fast reactors.

G_I does not measure the reactivity change directly. In the same way as G it represents the effect on reactivity that would result if the changes in composition of the fuel in the core were distributed uniformly throughout the core, whereas in fact the changes are greatest at the centre of the core where the flux is highest. In spite of this it is a useful guide to reactivity changes and when G_I is greater (i.e. less negative) the loss of reactivity with burnup is less.

It is important that the rate of change of reactivity with burnup should be as small as possible. The fuel is changed in batches when the reactor is shut down, and it is desirable to make the interval between changes as long as possible to minimise interruption to operation. Between changes the reactivity of the fuel declines and has to be compensated for by withdrawing control rods containing neutron absorbers from the core. When the fuel is fresh, therefore, neutrons that could have been used for breeding are absorbed in the control rods. The higher the internal breeding the smaller the reactivity loss with burnup, the smaller the quantity of neutron-absorbing material in the core and the better the overall breeding or, alternatively, the longer the period between fuel changes.

If a reactor core is designed to consume fissile material rather than breed G_I is substantially negative and the loss of reactivity with burnup is large. This can be a major economic and operational disadvantage for a consumer reactor.

1.4.4 Fuel Composition

As the fuel is irradiated in the reactor its isotopic composition changes. The effect is illustrated in Figure 1.17 which shows what would happen, in theory, to 1 kg of ^{238}U if it could remain in a reactor until it was entirely consumed by fission (something that is impossible in practice because the buildup of fission products would completely disrupt the fuel material). The quantity of ^{239}Pu would increase, followed by ^{240}Pu, then ^{241}Pu and lastly by ^{242}Pu, all in successively smaller amounts. On the linear scale of Figure 1.17 ^{243}Pu and all the americium and curium isotopes do not appear because the quantities produced are very small. This reflects the fact that, as far as any effect on the operation of the reactor is concerned, they are quite unimportant. Their effect on the waste stream, which is far from unimportant, is discussed later.

The extent of irradiation is indicated in Figure 1.17 in terms of the burnup, the fraction of the "heavy atoms" (i.e. nuclei of uranium or plutonium) that have been fissioned. Since all fissions, of whatever isotope, liberate roughly the same amount of energy (about 200 MeV), burnup can also be measured in terms of the energy released. 100% burnup – i.e. complete destruction by fission – would yield about 80 TJ/kg. Another convenient way of measuring burnup is in terms

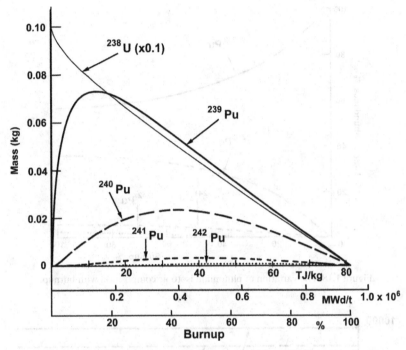

Figure 1.17 Complete burnup of 1 kg of ^{238}U.

of megawatt-days per tonne (MWd/t). Complete burnup is equivalent to about 10^6 MWd/t.

Figure 1.18 shows how the isotopic composition of the plutonium in Figure 1.17 changes with burnup. The most notable feature is the steady increase of the ^{240}Pu fraction. The concentrations of ^{241}Pu and ^{242}Pu are always small. The composition tends asymptotically towards 239:240:241:242 $\approx$ 56:36:5:2. (The precise ratios depend on the spectrum of the neutron flux.)

Figure 1.19 shows how the quantities of the various isotopes would vary with burnup in an idealised uranium-cycle reactor with a continuous feed of fertile ^{238}U. The reactor is assumed to generate 2500 MW (thermal) from a core containing a total of 7.4 tonnes of fissile and fertile material (i.e. heavy atoms). The initial enrichment in ^{239}Pu is

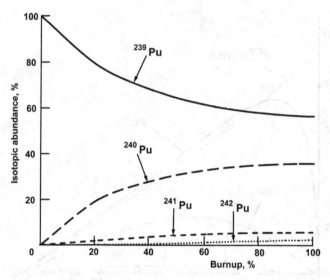

Figure 1.18 The variation of plutonium isotopic composition with burnup.

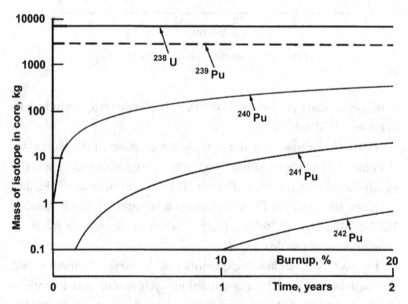

Figure 1.19 The evolution of fuel composition in a uranium-cycle reactor (2500 MW thermal, 7.4 tonnes of heavy atoms).

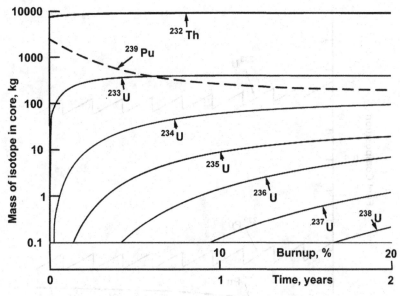

Figure 1.20 The evolution of fuel composition in a thorium-cycle reactor (2500 MW thermal, 7.4 tonnes of heavy atoms).

23% but this declines as the concentration of fissile ^{241}Pu builds up. The buildup of the other plutonium isotopes can also be seen.

Figure 1.20 shows the same information for a thorium-cycle reactor that has an initial loading of ^{239}Pu to make it critical. The uranium isotopes build up very slowly and for a prolonged period ^{239}Pu has to be added to maintain criticality. The fact that only very small quantities of ^{237}U and ^{238}U, the sources of higher actinides (see Figures 1.14 and 1.15), are produced indicates that thorium-cycle reactors produce less of the hazardous higher-actinide waste than uranium-cycle reactors.

However, as stated earlier, Figures 1.19 and 1.20 are theoretical because in practice the fertile feed is not continuous but comes in batches when the core is reloaded after irradiated fuel is removed for reprocessing. The maximum burnup achievable in practice is around 20%, at which point in a uranium-cycle reactor the plutonium contains around 77% ^{239}Pu and 21% ^{240}Pu. If plutonium with a higher ^{239}Pu

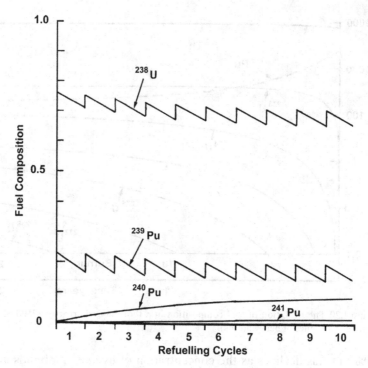

Figure 1.21 Changes in fuel composition in an operating reactor.

concentration is required the fuel has to be removed and reprocessed much sooner, after at most a few percent burnup. Plutonium rich in ^{239}Pu is sometimes called "high-grade plutonium".

Figure 1.21 shows how the isotopic abundances vary typically in a reactor. While it operates the quantities of both ^{238}U and ^{239}Pu decrease (usually they are said to "burn down"), the reactivity being maintained by withdrawing control absorbers from the core. At the end of a period of operation (usually called a "run") some of the irradiated fuel is removed and taken for reprocessing. The plutonium is separated, some fresh plutonium is added to it to replace that consumed, and it is returned to the reactor. In the example of Figure 1.21 it is assumed that the plutonium in the core at startup and added at each reload is pure ^{239}Pu. The quantities of the higher plutonium isotopes increase steadily from run to run. ^{242}Pu is present but its abundance is indistinguishable on the scale of Figure 1.21.

1.5 CONTROL RODS

1.5.1 Materials

The reactor is controlled and shut down by moving control rods incorporating neutron-absorbing material into or out of the core. Other methods of control such as moving fuel or parts of a neutron reflector around the core were used in early experimental reactors but are not feasible in a large power reactor.

The absorbing material is usually boron, possibly enriched in ^{10}B, in the form of the carbide B_4C. Alternatives are metallic tantalum or oxides of europium or gadolinium. ^{10}B has a very high capture cross-section that varies with energy E as $1/\sqrt{E}$ up to 0.1 MeV and thus captures neutrons predominantly at the low-energy end of the spectrum.

A disadvantage of boron is that it captures neutrons by an (n,α) reaction so that while it is in an operating reactor helium atoms accumulate within the crystals of the boron carbide. These tend to form little bubbles of gas that disrupt the structure of the crystals and damage the material. This, together with loss of the ^{10}B, limits the life of a rod used for controlling the reactor while it is operating. Shut-off rods, used only to shut the reactor down and hold it subcritical during refuelling, are not subject to this limitation.

1.5.2 Reactivity Worth

If a control rod is inserted a distance x into the core, as shown in Figure 1.22, the change in reactivity (neglecting any effects other than neutron capture) is proportional to the integral of $\phi\phi^*$ over the length of the rod times the difference between the capture cross-section of

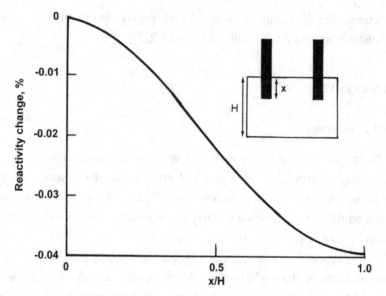

Figure 1.22 The reactivity of partially-inserted control rods.

the material making up the rod and that of the material (presumably coolant) it displaces – i.e.

$$\Delta\rho \propto \int_0^x \sum_g \phi_g \phi_g^* \left(\Sigma_{c,rod} - \Sigma_{c,coolant}\right) dx. \qquad (1.46)$$

The result is an S-shaped curve of reactivity against position as shown in Figure 1.22. The reactivity change on inserting a control rod fully into the core is usually called the "worth" of the rod.

This first-order perturbation theory estimate is not an accurate indication of the worth because the distributions of ϕ and ϕ^* are altered by the presence of the rod. Figure 1.23 shows the axial variation of the total flux ϕ with control rods withdrawn and with all the control rods inserted a third of the way into the core. The presence of the rods in the top of the core pushes the flux towards the bottom.

Figure 1.24 gives an impression of the flux distribution around a single partially-inserted rod and indicates how the flux is depressed in its vicinity. The result is that the reactivity worth of one control rod

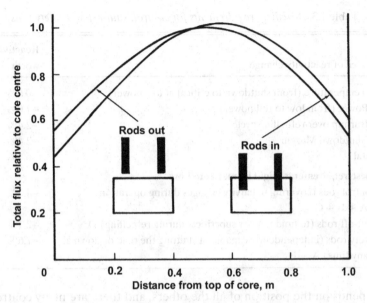

Figure 1.23 Displacement of the neutron flux by partially-inserted control rods.

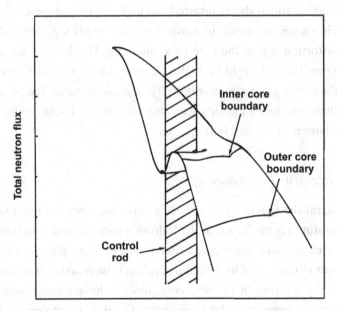

Figure 1.24 Flux depression in the vicinity of a control rod.

Table 1.3 *Reactivity requirements for control, shut-off and safety rods*

Source of reactivity change	Reactivity worth
A Temperature (from shutdown to critical at low power)	−0.6%
B Power (from low to full power)	−1.0%
C Burnup over refuelling cycle	−2.4%
D Shutdown Margin	−2.4%
Total	−6.4%
These requirements might be met as follows	
Control rods (covering reactivity changes during operation) A + B + C	−4.0%
Shut-off rods (to hold reactor subcritical during refuelling) D	−2.4%
Safety rods (independently capable of shutting the reactor down at any time) A + B	−1.6%

depends on the position of all the others, and there are many control rods and shut-off rods in a large reactor. The worth of a rod is lower if the neighbouring rods are inserted and higher if they are not.

It is normal, however, to move the control rods together to keep flux distortion across the core to a minimum. If this were not done the power density might be higher on one side of the core than the other, causing greater non-uniformity of coolant outlet temperature and therefore more entropy gain due to mixing. There would also be variations of the fuel burnup rate.

1.5.3 Reactivity Requirements

In general the reactivity of a reactor core decreases with increasing temperature (as explained in the following section) and with burnup. Thus a new, cold reactor core is in its most reactive state and a core at power at the end of its refuelling cycle is least reactive. The control rods, fully inserted in the new core, have to have enough negative reactivity to compensate for these changes so that, when they are fully withdrawn at the end of the cycle, the reactor remains critical. Table 1.3 gives typical values of the reactivity requirements for a breeder reactor.

The "shutdown margin" is a safety allowance to make sure that the reactor remains subcritical during refuelling even if errors are made and excessive fuel is added. It also has to allow for one or more control rods themselves being removed and replaced. The safety rods are there so that the reactor can be shut down from any operating condition even if some of the control rod mechanisms fail to work, including the possibility of a rod being accidentally withdrawn. These points are discussed in more detail in Chapter 5.

In principle there is no difference between the control, shut-off and safety rods: they could be identical items distinguished only by different names to denote their different functions. However as explained in Chapter 5 the reliability of the shutdown system, which is crucial for the safety of the reactor, is enhanced by diversity. Thus the shut-off rods might utilise a different absorber material from the control rods or different operating mechanisms.

It is important to note that Table 1.3 applies to a typical breeder reactor in which the loss of reactivity by burnup of fissile nuclides is reduced substantially by internal breeding of new fissile material in the core (see section 1.4.3). In the core of a reactor designed to consume fissile material the internal breeding is much reduced or possibly eliminated entirely, so that the loss of reactivity with burnup is much greater, possibly three times as much as indicated in Table 1.3. In a typical breeder reactor core some 7–10% of the space is occupied by control rods (including shutdown and safety rods), but in a consumer core the proportion is likely to be greater. Loss of reactivity with burnup may in fact limit the length of the refuelling cycles for such a reactor.

1.6 REACTIVITY COEFFICIENTS

1.6.1 Effects of Temperature

Temperature affects reactivity in a number of ways. The temperature of the structure affects the dimensions of the reactor core and sometimes the relative position of the various parts: the densities of all the materials depend on temperature (but the most important effects arise from changes in the density of the coolant), and the temperature of the fuel affects the resonance self-shielding in the fuel materials. The various effects are discussed in detail by Hummel and Okrent (1970).

If the temperature changes are small the resulting changes in dimensions, density and self-shielding are also small in most cases and proportional to the temperature changes. First-order perturbation theory is valid and the resulting reactivity changes are, approximately at least, linear and independent. As a result it is useful to express temperature-induced reactivity changes as reactivity coefficients of the form $d\rho/dT_i$, where T_i is the temperature in question (the coolant inlet temperature, for example, or the mean fuel temperature) and $d\rho/dT_i$ can be taken to be constant and independent of all the T_i.

This approximation breaks down in some cases. It may not be true in normal operation if there is intermittent contact between bowing fuel elements, and it is certainly untrue in the extreme conditions that may be encountered in an accident – if the coolant boils, for example, or if the fuel temperature rises very high.

1.6.2 Structure Temperatures

The effects of the temperatures of the many different parts of the reactor structure on reactivity depend on the detailed design of the reactor. The possible overall effects can be illustrated by the following examples, but the reality may be much more complex.

The radial dimensions of the reactor core are determined by the temperature of the structure that supports it while the axial dimensions of the core may depend mainly on the temperature of the fuel cladding, so radial and axial dimensions may change independently with temperature. If the structural temperatures increase the mean "smeared" densities of the solid materials decrease, but the coolant mean density may remain the same or it may actually increase. If, for

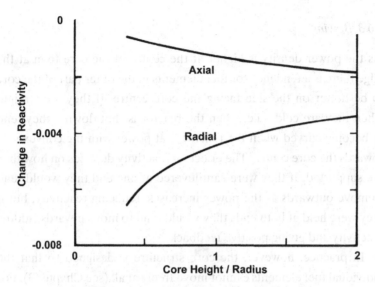

Figure 1.25 The effect of 1% linear expansion on reactivity.

example, the temperatures of all the materials in the core are constant while the temperature of the structure that supports it increases so that its radius increases by 1%, the core volume will increase by 2%. The actual volume and mass of fuel and structural material in the core will remain the same, so their "smeared" densities will decrease by 2%. If the fuel and structure made up 50% of the original core, however, they will now form only 49% of the expanded core, so that the coolant volume fraction will have increased from 50 to 51%.

Figure 1.25 shows the effects of 1% increases in the radial and axial dimensions of a typical cylindrical oxide-fuelled breeder core as a function of the ratio of height to radius, H/R. In all cases the reactivity decreases with expansion because the leakage increases (in simple terms the "gaps" between the atoms are greater so the neutrons are more likely to diffuse out), but if H/R is small the effect of increased height is small. In principle, in the limit of an infinite slab reactor ($R \to \infty$), uniform expansion in the axial direction has no effect on reactivity at all.

1.6.3 Bowing

As the power density is higher at the centre of the core than at the edge there is a tendency for fuel elements in the outer part of the core to be hotter on the side facing the core centre. If they are straight when they are cold – i.e. when the reactor is shut down – they tend to become curved when the reactor is at power with the convex side towards the core centre. The effect on reactivity depends on how they are supported. If they were cantilevered at one end they would tend to move outwards as the power increases, reducing reactivity, but if they were held at both ends they would tend to move inwards, adding reactivity and giving positive feedback.

In practice, however, the core structure is designed so that the individual fuel elements cannot move freely at all (see Chapter 3). For this reason the effect of bowing on reactivity is likely to be small, but there may be complicated nonlinear effects as fuel elements or sub-assemblies distort to take up small clearances within the manufacturing tolerances.

1.6.4 Coolant Density

Equation 1.25 shows three effects of a change in density of a non-fissile isotope: a "moderating" effect due to the change in $\Sigma_{sg \rightarrow g'}$, a "capture" effect due to the change in Σ_{rg}, and a "scattering" effect due to the change in D_g. There is a fourth "self-shielding" effect, which is shown in equation 1.40: if Σ_s is changed all the group cross-sections for the resonant isotopes, such as Σ_{cg} in equation 1.38, are altered. For lead-bismuth or gas coolant all these effects are small but not so for sodium. The various components of the sodium temperature coefficient of reactivity are as follows.

An increase in temperature reduces the density of the sodium and this reduces Σ_{cg} giving an increase in reactivity and a positive capture contribution to the temperature coefficient, but it is a small contribution because the sodium does not capture many neutrons in the first place.

The reduction in $\Sigma_{sg \to g'}$ has the effect of hardening the spectrum – that is of shifting the peak in ϕ shown in Figure 1.7 to a slightly higher energy. Figure 1.7 also shows that in a plutonium-fuelled reactor the peak occurs in the range where ϕ^* increases with energy, so there is a gain in reactivity. Because sodium is an effective moderator this positive moderating contribution to the sodium temperature coefficient is large.

As Σ_s decreases Σ_{cg} decreases because ϕ_g in equation 1.38 increases more than J. There is a similar effect on Σ_f but, as in the case of a change in the fuel temperature explained later, in a breeder reactor the capture effect from ^{238}U is greater than the fission effect from ^{239}Pu and ^{241}Pu so the reactivity increases, giving another small positive contribution to the sodium temperature coefficient.

These three components, due to the capture, moderating and self-shielding effects, are contributions to the $\phi_g \phi_g^*$ term in equation 1.25, so they all depend on position in roughly the same way. They are greatest at the centre of the core and smaller at the edges. The effect of scattering is quite different, however. A decrease in the scattering cross-section increases the diffusion coefficient D_g and this results in a decrease in reactivity, depending this time on $\nabla \phi_g \cdot \nabla \phi_g^*$. It is therefore zero at the centre and reaches a maximum in the outer parts of the core.

Table 1.4 shows the effect of increasing the coolant temperature uniformly throughout the core of a small sodium-cooled breeder. In this case the resulting reactivity change is the small difference between two large quantities and is just positive, giving a small overall positive sodium temperature coefficient.

The effect of a local change in sodium temperature or of a local loss of sodium may, on the other hand, be positive or negative depending on where in the core it happens. Figure 1.26 shows the effect of loss of coolant from various points on the axis of the core of a larger breeder

Table 1.4 *Components of the sodium temperature coefficient of reactivity of a small fast reactor*

Component	$\partial k/\partial T$ (K^{-1})
Moderation	5.22×10^{-6}
Capture	0.78×10^{-6}
Self-shielding	0.52×10^{-6}
Scattering	-6.51×10^{-6}
Total	0.01×10^{-6}

reactor. Near the centre moderating, capture and self-shielding dominate and the reactivity change is positive, but towards the edges leakage becomes more important and it is negative.

Chapter 5 describes extreme hypothetical accidents in which the sodium is lost completely from all or part of the core. The reactivity

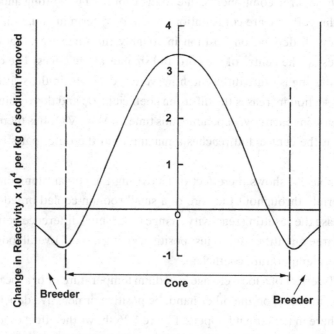

Figure 1.26 The spatial variation of sodium void reactivity.

increase due to complete loss of coolant from the whole core of the small reactor of Table 1.4 would be 4×10^{-5}, but if coolant was lost only from the part of the core where the sodium coefficient is positive it would be as high as 7×10^{-3}. For a large 2500 MW (heat) reactor the effects are more positive, and the gain in reactivity might be 0.017 for complete loss of sodium and 0.020 if it was lost only from the central region.

1.6.5 Doppler Coefficient

As the fuel temperature increases the resonance self-shielding in the fuel isotopes decreases. The resulting change in reactivity is called the Doppler effect because it is due to the dependence of the effective energy of a neutron on the relative motion of the nucleus with which it interacts. The decrease in self-shielding results in an increase in all the cross-sections of the fuel isotopes at energies below about 20 keV.

Although all the cross-sections increase, including scattering, the most significant changes are in Σ_c and Σ_f. In all breeder reactors the increase in capture in ^{238}U dominates over the increase in fission in ^{239}Pu and the Doppler effect is negative, but the same is not always true for consumer reactors.

The Doppler effect is not independent of temperature. The dependence of J (equation 1.39) on T is complex. If β/ψ_0, where ψ_0 is the peak value of ψ, is large (corresponding to a resonance with a low peak at higher energy) then $J \propto T^{-1/2}$ and $dJ/dT \propto T^{-3/2}$, but if β/ψ_0 is small $J \propto T$ and dJ/dT is constant. Thus the Doppler effect varies in a different way with temperature for different resonances, but the total coefficient, involving a weighted sum of dJ/dT for all the resonances, varies roughly as $1/T$. As a result it is convenient to define a "Doppler constant", $-T \, d\rho/dT$.

Figure 1.27 shows the contribution to the Doppler reactivity effect from various isotopes in different neutron energy groups for a breeder reactor with low enrichment. The most important effects are around

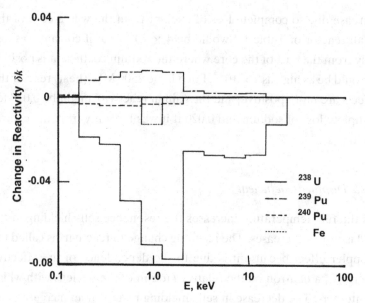

Figure 1.27 The distribution of the Doppler coefficient in energy.

1 keV at which energy the resonances are not resolved. The effect of
the iron resonance at 1 keV can be seen. This is the only resonance in
a non-fissile or fertile isotope that makes any significant contribution
to the Doppler effect.

In an operating reactor the fuel is not all at the same temperature,
and as the Doppler coefficient is a function of temperature there is
a question of the correct average temperature to use. In a cylindrical
fuel element with constant thermal conductivity the volume-weighted
mean temperature $\overline{T}$ is halfway between the central and surface temperatures. Because the Doppler coefficient is higher at low temperatures the effective temperature T_{eff} for the Doppler effect is slightly
lower than $\overline{T}$, but the difference is small. If the temperatures in the
fuel element are in equilibrium, it can be shown that

$$\frac{T_{eff}}{\overline{T}} \approx 1 - \frac{1}{3}\left(\frac{\overline{T} - T_0}{\overline{T}}\right)^2 , \tag{1.47}$$

where T_0 is the fuel surface temperature. Temperature variation across the core can be taken into account by weighting changes with $\phi_g\phi_g^*$ as indicated in equation 1.25.

1.6.6 Power and Temperature Coefficients

In normal operation the temperatures of the various materials in the core vary in a regular way and it is useful to sum up their combined effects. There are various ways of doing this, one of which is to determine the "isothermal temperature coefficient". This is the rate of change of reactivity with temperature if the temperatures throughout the core change by the same amount. It includes the effects of axial and radial expansion of the structure, expansion of the coolant, and the Doppler effect. Because of the variation of the Doppler coefficient it depends slightly on the fuel temperatures. For a typical sodium-cooled breeder it might be -2.5×10^{-5} K^{-1} when the fuel is cold (i.e. the reactor is operating at very low power), and -1.5×10^{-5} K^{-1} at normal operating temperature. It can be thought of as measuring the response of the reactor to a change in coolant inlet temperature when power and coolant flow-rate are held constant.

Another useful parameter is the power coefficient. This is the rate of change of reactivity with power, assuming the coolant flow-rate changes proportionately to the power and the inlet temperature is constant. The coolant and structure temperatures everywhere are thus constant so that the power coefficient depends only on the Doppler effect. A typical value of the power coefficient for a 2500 MW (heat) sodium-cooled breeder is -3×10^{-6} (MW)$^{-1}$.

1.6.7 Dependence of Doppler and Sodium Coefficients on Design – Breeders

As explained in Chapter 5 the behaviour of the reactor in a serious accident may be strongly dependent on both the Doppler effect and the reactivity change due to loss of coolant. The consequences of accidents are likely to be less severe the more negative these are. It is important therefore to know how design changes affect them.

An important factor is the "enrichment", the ratio of fissile to fertile material in the core fuel. Figure 1.10 illustrates this. For the small highly-enriched core the increase of ϕ^* with neutron energy is much lower than for the larger core because of the reduced probability of ^{238}U fission and because high energy neutrons are more likely to leak out of the core. As a result, in a sodium-cooled breeder, the moderating component of the sodium coefficient becomes less positive as the enrichment is increased, and at the same time the Doppler coefficient is made less negative by the reduced amount of ^{238}U.

It is not necessary to make the core smaller to increase the enrichment. The height of the core can be reduced or regions or layers of breeder can be incorporated within the core. Changes of this type make the sodium coefficient more negative, the Doppler coefficient less negative, the breeding ratio higher and the critical mass larger. The change in sodium coefficient is probably more important than the reduction in the Doppler coefficient from the point of view of safety and reduced-height cores may be attractive as a result. The increase in critical mass outweighs the improved breeding, however, and together with the increased mechanical complexity there is an economic disadvantage. Figure 1.28 shows the effect of changing the height of the cylindrical core of a sodium-cooled breeder while keeping the volume constant.

Another way to reduce the sodium coefficient is to make the spectrum softer by incorporating more moderating material in the core, so that the effect of losing the moderation by the sodium is less important. Solid moderators such as beryllium oxide, BeO, or boron carbide depleted in ^{10}B (i.e. ^{11}B$_4$C) have been proposed. If some 10% of the core volume were to be occupied by BeO the sodium coefficient could be made negative and in addition the Doppler coefficient would be more negative. This is in fact the only practical way that both

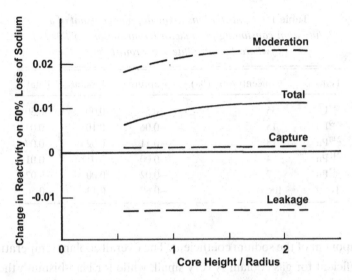

Figure 1.28 The effect of core shape on sodium-loss reactivity.

Doppler and sodium coefficients can be made more negative at the same time, but there is a disadvantage in that the critical mass is increased.

Increasing the fraction of ^{241}Pu in the fuel also makes both Doppler and sodium coefficients more negative, but this is not really a useful design option because the effects are small and it is likely to be difficult to alter the composition of the plutonium in the fuel without a severe cost penalty. The effect on the sodium coefficient arises because the fission cross-section of ^{241}Pu increases more rapidly with decreasing energy than that of ^{239}Pu, so the importance of low-energy neutrons is increased. The higher fission cross-section of ^{241}Pu in general means that more fertile material (^{238}U + ^{240}Pu) is present and the Doppler coefficient is more negative.

Replacing oxide fuel by carbide makes the spectrum harder and reduces the enrichment so the sodium coefficient becomes more positive and the Doppler coefficient more negative.

Since neither lead-bismuth nor gas has significant moderating effects the coolant temperature coefficients for both lack the positive

Table 1.5 *Contributions to the doppler constant of a*
plutonium-consuming fast reactor (contributions to $T \partial k / \partial T$
where $T =$ absolute temperature, %)

Isotope	Concentration (%)	Capture	Fission	Total
^{238}U	55	−0.50	0.00	−0.50
^{239}Pu	15	−0.06	0.10	0.04
^{240}Pu	18	−0.11	0.02	−0.09
^{241}Pu	4	−0.00	0.01	0.01
^{242}Pu	7	−0.02	0.00	−0.02
Total	100%	−0.69	0.13	−0.56

component of the sodium coefficient. The overall coolant temperature coefficient for gas coolant is very small, while for lead-bismuth there is a small negative leakage effect.

Figure 1.11 shows that the effect of a change in the sodium density is very different for a thorium-cycle breeder. There is little variation of ϕ^* with energy so the moderating component of the sodium coefficient is much smaller. At the same time the Doppler coefficient is strongly negative because of the high importance of neutrons in the 1 keV energy range.

1.6.8 Dependence of Doppler and Sodium Coefficients on Design – Consumers

A reactor designed to consume plutonium rather than breed it inevitably has less ^{238}U in the core, so the negative component of the Doppler coefficient is reduced, but it is not lost. Table 1.5 shows the various components of the Doppler effect of losing all the sodium from the fuel subassemblies in a 2500 MW (thermal) core with fuel enriched to 45%. The plutonium is assumed to have been recycled once in a thermal reactor. The enrichment is the maximum for which $(U,Pu)O_2$ is soluble for reprocessing.

It will be seen that even with this low concentration the negative effect of capture in ^{238}U dominates the Doppler coefficient. Moreover even if there were no ^{238}U at all the coefficient, for this highly irradiated plutonium, would still be slightly negative mainly because of the contribution of the ^{240}Pu. However for "high-grade" plutonium rich in ^{239}Pu it would be positive. As in the case of a breeder the Doppler coefficient can be made more negative by incorporating solid moderating material in the core.

In a consumer core compared with a breeder the reduced effect of fission in ^{238}U makes the neutron importance spectrum flatter and thus the moderation component of the sodium coefficient smaller. However the effect of increasing the concentrations of the higher plutonium isotopes in a consumer reactor is the opposite of that in a breeder. Although in both case the fertile-to-fissile ratio increases the fact that the fast-neutron fission cross-section of ^{240}Pu is higher than that of ^{238}U makes the importance spectrum steeper and the moderating component larger.

1.7 SUBCRITICAL REACTORS

1.7.1 Neutron Economy

To ensure the safety of a reactor it is essential to control the reactivity in such a way that it does not become supercritical and generate excessive amounts of heat. This may be easier if the reactor is designed to remain subcritical, so that criticality is not possible at least in normal operation.

A subcritical reactor is a multiplying assembly with $k_e < 1$ that is driven by a neutron source. The relationship between power P, source-strength S and k_e is of the form

$$P \propto S/(1 - k_e). \tag{1.48}$$

The neutrons might come from spontaneous fission of curium isotopes (see section 1.4.1), or the $^9Be(\alpha,n)^{12}C$ reactions in a radium-beryllium assembly, but for a power reactor a controllable source is required, such as can be provided by a particle accelerator. Such a system is known as an "accelerator-driven reactor" (ADR) or "accelerator-driven system" (ADS). In most cases it involves an accelerator (a linear accelerator for example or a cyclotron) that delivers a beam of high-energy protons to a spallation target consisting of material with high atomic weight located in the subcritical core. The target produces a cascade of high-energy neutrons that drive the reactor. An ADR can be seen as a device for producing neutrons that can be used for a variety of purposes – breeding fissile material such as ^{233}U from thorium, transmuting radioactive nuclear waste (either fission products or higher actinides) to reduce its hazard, or generating power.

The subcritical assembly acts to multiply the neutron output of the spallation source, and the multiplying factor depends on its k_e. Figure 1.29 shows the neutron economy of an ADR. It is greatly simplified but illustrates the important aspects.

This simplified analysis shows that $\eta + S = 1 + L + C$, where, for each fissile nucleus destroyed, $\eta = \bar{v}/(1 + \alpha)$, and $\alpha = \sigma_f/(\sigma_f + \sigma_c)$ is the number of fission neutrons generated; S is the number of source neutrons injected into the assembly; L is the number of neutrons lost by being captured in structure, coolant, shielding, etc.; and C is the number of neutrons available for use. In addition since the chain reaction produces η new neutrons from $\eta + S$ neutrons, $k_e = \eta/(\eta + S)$, so, after rearranging the terms,

$$C = \frac{S}{(1 - k_e)} - 1 - L. \tag{1.49}$$

For a critical reactor $S = 0$ and $k_e = 1$, and

$$C = \eta - 1 - L. \tag{1.50}$$

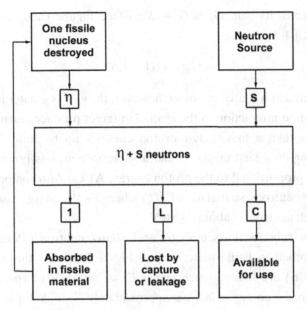

Figure 1.29 The neutron economy of an accelerator-driven reactor.

If the ADR is operated as a breeder, C is the breeding ratio. If it is operated as a consumer of higher actinides account must be taken of the capture events in the fissile fuel, each of which may lead to the production of a new higher actinide nucleus, so the net reduction in the number of higher actinide nuclei is $C - \sigma_c/(\sigma_c + \sigma_f) = C\alpha/(1 + \alpha)$ per fuel atom destroyed.

As explained in the Introduction for a critical reactor L is in practice around 0.2, mainly because neutrons are captured in control absorbers. However it may be possible to control an ADR by means of varying the neutron source strength, in which case control rods may be unnecessary. This might allow L to be reduced to around 0.1.

1.7.2 Gain

There are two ways in which an ADR can be considered as an amplifier, augmenting the output of the accelerator. Considered as a source of neutrons, its gain G_n is $(\eta + S)/S$ (see Figure 1.29), and since $k_e = \eta/(\eta + S)$,

$$G_n = 1/(1 - k_e). \tag{1.51}$$

Considered as a source of power, however, the gain is greater because each proton interaction in the spallation target produces many more neutrons than a fission. For proton energies up to about 1 GeV impinging on a lead target n_n, the number of neutrons produced, is roughly proportional to the proton energy. A1 GeV proton produces about 18 neutrons, so that $n_n \approx E_p/E_0$ where E_0, the energy associated with each neutron, is about 55 MeV.

In the subcritical assembly for each source neutron $1/S$ neutrons are absorbed in fissile material (see Figure 1.29). Of these events $1/S(1 + \alpha)$ are fissions, each giving $E_f \approx 200$ MeV of energy output. Thus the energy gain G_e is approximately $E_f/E_0 S(1 + \alpha)$. Since $S = \eta(1 - k_e)/k_e$ and $\eta(1 + \alpha) = \bar{v}$,

$$G_e = \frac{E_f k_e}{E_0 \bar{v}(1 - k_e)}. \tag{1.52}$$

Since k_e in the numerator can be taken to be ≈ 1, and with $E_f/E_0 \approx 200/55 = 3.6$,

$$G_e \approx \frac{3.6}{(1 - k_e)\bar{v}}. \tag{1.53}$$

The numerator in this expression depends on the nature of the spallation target and $\bar{v}$ depends on the fissile material.

1.7.3 Changes in Reactivity

As the composition of the fuel changes with burnup the reactivity changes. If the power is to be kept constant equation 1.48 shows that either the source strength has to be changed by altering the accelerator beam current, or the reactivity has to be adjusted by moving control rods. However one of the potential advantages of an ADR over a critical reactor is the possibility of dispensing with control rods altogether. This reduces L, making more neutrons available for use (equation 1.49), and it also reduces the complexity and therefore the cost of the plant.

If the ADR has no control rods it has to remain subcritical, with a safety margin, at all times in the refuelling cycle, and this means that at the times in the cycle when the fuel is least reactive k_e is significantly less than one. This has to be compensated for by increasing the beam current. If, for example, the reactivity changes by 3% during a run, with a 1% safety margin k_e varies between 0.96 and 0.99, which requires the beam current to be changed by a factor of 4 to keep the power constant.

This imposes limitations on the design and cost of the accelerator, which has to be powerful enough to provide the high current needed when k_e is low but is operating well below its capacity most of the time. To avoid this disadvantage, therefore, it may be preferable to accept the alternative disadvantage of installing control rods. It may then be possible to operate the plant close to critical – possibly with $k_e \approx 0.995$ – at all times, with a steady beam current and steady power.

1.7.4 Power Density

The neutron flux distribution in an ADR is affected by the degree of subcriticality. Figure 1.30 shows the radial distribution of the total flux in ADRs with $k_e = 0.995$ and 0.95 compared with that in a critical reactor. The core is cylindrical, 1 m high and 2 m in diameter, with properties similar to those of the standard sodium-cooled oxide-fuelled core described in Table 1.1, but with a neutron source located on the axis. It will be seen that the source has little effect on the flux in the outer part of the core, but is higher close to the source and more so the greater the subcriticality. As in a critical reactor the effect on the power distribution can be reduced by varying the composition of the fuel in different radial zones (see section 1.3.2), but the peak close

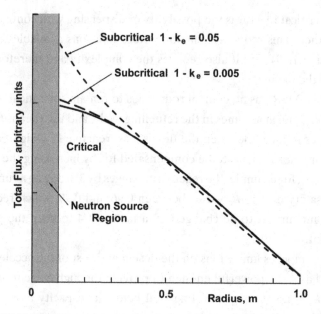

Figure 1.30 The effect of subcritical reactivity on the flux distribution in an ADR.

to the source may necessitate specific design measures to avoid local overheating.

REFERENCES FOR CHAPTER 1

Baker, A. R. and R. W. Ross (1963) Comparison of the Values of Plutonium and Uranium Isotopes in Fast Reactors, pp 329–265 in *Breeding, Economics and Safety in Large Fast Power Reactors* Report ANL 6792, USAEC, Washington

Broomfield, A. M., C. F. George, G. Ingram, D. Jakeman and J. E. Sanders (1969) Measurements of k-infinity, Reaction Rates and Spectra in ZEBRA Plutonium Lattices, pp 1502–1511 in *Fast Reactor Safety Technology, Volume 3*, American Nuclear Society, LaGrange Park, Illinois, USA

Brown, F. B. (2012) *Fundamentals of Monte Carlo Particle Transport* Report LA-UR-05-4983, Los Alamos National Laboratory, Los Alamos, New Mexico, USA

Coates, D. J. and G. T. Parks (2010) Actinide Evolution and Equilibrium in Fast Thorium Reactors, *Annals of Nuclear Energy*, 37, 1076–1088

Davison, B. and J. B. Sykes (1957) *Neutron Transport Theory*, Clarendon, Oxford

Duderstadt, J. J. and L. J. Hamilton (1976) *Nuclear Reactor Analysis*, Wiley, New York

Greenspan, H., C. N. Kelber and D. Okrent (1968) *Computing Methods in Reactor Physics*, Gordon and Breach, New York

Hummel, H. H. and D. Okrent (1970) *Reactivity Coefficients in Large Fast Power Reactors*, American Nuclear Society, Hinsdale, Illinois, USA

Okrent, D. (1961) Performance of Large Fast Power Reactors Including Effects of Higher Isotopes, Recycling and Fission Products, pp 271–297 in *Physics of Fast and Intermediate Reactors, Volume 2*, IAEA, Vienna

Okrent, D., K. P. Cohen and W. B. Lowenetein (1964) Some Nuclear and Safety Considerations in the Design of Large Fast Power Reactors, pp 147–148 in *Peaceful Uses of Atomic Energy, Volume 6*, United Nations, New York

Palmiotti, G., E. E. Lewis and C. B. Carrico (1995) *VARIANT: Variational Anisotropic Nodal Transport for Multidimensional Cartesian and Hexagonal Geometry Calculation* Report ANL-95/40, Argonne National Laboratory, Argonne, Illinois, USA

Tamplin, L. J. (Ed) (1963) *Reactor Physics Constants* Report ANL 5800 (2nd ed.), USAEC, Washington, DC

Van der Meer, K. et al. (2004) Spallation Yields of Neutrons Produced in Thick Lead/Bismuth Targets by Protons at Incident Energies of 420 and 590 MeV, *Nuclear Instruments and Methods in Physics Research B*, 217, 202–220

Wardleworth, D. and R. C. Wheeler (1974) Reactor Physics Calculational Methods in Support of the Prototype Fast Reactor, *Journal of the British Nuclear Energy Society*, 13, 383–390

Yiftah, S. (1961) Effect of the Plutonium Isotopic Composition on the Performance of Fast Reactors, pp 257–270 in *Physics of Fast and Intermediate Reactors, Volume 2*, IAEA, Vienna

2

FUEL

2.1 INTRODUCTION

In common with pressurised water reactors (PWRs), boiling water reactors (BWRs) and advanced gas-cooled reactors (AGRs) most fast power reactors use oxide fuel. There is a certain amount of experience with metal fuel in the United States, and there is interest in the use of metal fuel and other ceramic fuels such as nitride or carbide for future fast reactors. This chapter deals mainly with the oxide and metal fuels of which there is most experience, and it covers other fuel materials in less detail.

Since oxide fuels are widely used much of this chapter applies as well to thermal reactors as to fast reactors. The main difference is that new fast reactor fuel usually consists of a mixture of plutonium and uranium dioxides whereas at present most thermal reactors use enriched uranium dioxide with a small amount of plutonium present after irradiation. The use of mixed uranium and plutonium dioxide (MOX) fuel in thermal reactors is increasing. Most water-cooled thermal reactors have fuel clad in zirconium alloy, but AGRs use stainless steel, and because of the similarity of the coolant temperatures there is much in common between the behaviour of AGR fuel and that of a sodium-cooled fast reactor.

There is an important difference, however. In all systems it is economically desirable to irradiate the fuel as long as possible before it is removed from the reactor. In a thermal reactor the limit is set by the loss of reactivity as fission products that absorb thermal neutrons accumulate. In a fast reactor the loss of reactivity is much smaller, especially if the internal breeding is high, so the irradiation limit is different. It is set by the need to be sure that the cladding will remain intact and not allow radioactive material to escape into the coolant. As a result the maximum burnup is very high in a fast reactor. The neutron irradiation accompanying this burnup damages the cladding and the irradiation limit is set by the need to remove the fuel before the cladding loses its integrity.

A very thorough account of reactor fuel in general, and in particular of oxide fuel for fast reactors, is given by Olander (1976). It is particularly valuable for its treatment of the theory of many aspects of fuel behaviour.

2.2 OXIDE FUEL TEMPERATURES

2.2.1 Temperature Distribution

In a cylindrically symmetrical fuel element, the temperature T_{fs} of the surface of the fuel is related to the temperature of the surrounding coolant T_c by

$$T_{fs} - T_c = \frac{q}{2\pi} \left(\frac{1}{R_s h} + \frac{\ln(R_s/R_f)}{K_s} + \frac{1}{R_f h_f} \right), \qquad (2.1)$$

where q is the linear heat rating in the fuel element, h is the heat transfer coefficient between cladding and coolant, K_s is the thermal conductivity of the cladding, and h_f is the heat transfer coefficient between the fuel and the inside surface of the cladding. R_s is the outer radius of the cladding and R_f is its inner radius, which we can also take to be the outer radius of the fuel because, as we shall see, at times the fuel is in close contact with the cladding, and even when it is not the gap between them is small.

In a fast reactor there is usually no significant flux depression in the fuel element and the fission rate can be taken to be approximately uniform over its cross-section. The temperature distribution within the fuel itself however is complicated by the fact that over the wide temperature range in question the variation of the fuel conductivity K_f has to be taken into account. If the fuel temperature at a distance r from the axis is $T_f(r)$, and if the fuel is solid and heat is generated uniformly throughout it, then

$$\int_{T_{fs}}^{T_f(r)} K_f(T)dT = \frac{q}{4\pi}\left(1 - \frac{r^2}{R_f^2}\right),\qquad (2.2)$$

and the maximum fuel temperature T_{fm} is given by

$$\int_{T_{fs}}^{T_{fm}} K_f(T)dT = \frac{q}{4\pi}.\qquad (2.3)$$

Under some circumstances, however, there is a cylindrical hole in the centre of the fuel. If this has a radius R_h and heat is generated uniformly throughout the annular fuel region the temperature at r is given by

$$\int_{T_{fs}}^{T_f(r)} K_f(T)dT = \frac{q}{4\pi(1-\alpha)}\left(1 - \frac{r^2}{R_f^2} + 2\alpha \ln\left(\frac{r}{R_f}\right)\right),\qquad (2.4)$$

where $\alpha \equiv R_h^2/R_f^2$. The maximum temperature is given by

$$\int_{T_{fs}}^{T_{fm}} K_f(T)dT = \frac{q}{4\pi}\left(1 + \frac{\alpha \ln \alpha}{(1-\alpha)}\right).\qquad (2.5)$$

Comparison of equations 2.3 and 2.5 shows that for constant q the presence of a central hole reduces the maximum temperature.

2.2.2 Thermal Conductivity

At low temperature heat is conducted mainly by the diffusion of phonons through the crystals. As the temperature increases the density of phonons increases and, as the behaviour of the crystal lattice is slightly nonlinear, the probability of interaction between phonons increases, their mean-free-path decreases and the conductivity decreases. At higher temperature still electrons become more mobile and make a growing contribution to the conductivity.

For mixed oxide the conductivity is affected by the chemical composition, which can be described as $(Pu_aU_{(1-a)})O_{2+x}$, where a represents the plutonium fraction and x, which can be either positive or negative, the departure from the stoichiometric oxygen content. It is usually in the range $-0.03 < x < 0.03$.

The effect of stoichiometry on conductivity is as follows. If there is no plutonium ($a = 0$), below about 1400 °C K_f decreases steadily as x increases, but above 1600 °C K_f is a maximum for $x = 0$ and is reduced by departure from the stoichiometric composition in either direction. This seems to be because at high temperature an excess or a deficiency of oxygen, not exceeding a few per cent, is accommodated by vacancies or interstitials in the lattice. These defects reduce the phonon mean-free-path and hence the conductivity. At low temperatures, however, excess uranium ($x < 0$) is precipitated as the metal and the resulting free electrons increase the conductivity.

The situation is quite different if plutonium is present. Uranium and plutonium have nearly the same ionic radius and stoichiometric UO_2 and PuO_2 form solid solutions in whatever ratio they are mixed (for all values of a). But if the mixed oxide is not stoichiometric ($x < 0$) the excess metal is accommodated by means of lattice defects for $x > -0.02$ and by precipitation of Pu_2O_3 for $x < -0.02$. Thus for mixed oxide with a Pu fraction typical of a large fast reactor ($a \sim 0.2$) the conductivity decreases as the composition departs from stoichiometric in either direction ($x < 0$ and $x > 0$) at all temperatures. The effect is quite large: for $x = +/-0.02$, K_f is reduced to about 75% of its value for $x = 0$. For $x = 0$ the conductivity decreases slowly as a increases, so that in the range 400–1200 °C, for $a = 0.2$ K_f is about 13% smaller than for $a = 0$.

This picture is confused in practice by structural effects that can be much more important than the effects of composition. Principal among these is the effect of porosity, which reduces the effective conductivity. As explained later (section 2.3.2) the fuel may be manufactured with 10% or more porosity (i.e. 10% of the overall volume of the fuel may be occupied by voids). Porosity of 10% can reduce the effective thermal conductivity by as much as 25%.

Estimation of the effect of porosity in an operating fuel element is made very difficult because, as explained in section 2.4.1, the pores move under the influence of the temperature distribution, so that the conductivity changes with time and in different ways in different parts of the fuel. Further confusion is provided by the cracks in the fuel that are formed under the influence of differential thermal expansion and that have a large but unpredictable effect on conduction. Finally fission products, which themselves move through the fuel (section 2.4.6), and changes in crystal structure (section 2.4.1) affect conductivity in a manner that is not accurately known.

The maximum linear heat rating q is set by the requirement that the fuel should not melt. The melting points of UO_2 and PuO_2 are about 2850 °C and 2430 °C respectively, and the *solidus* temperature for $a = 0.2$ is about 2730 °C. Figure 2.1 shows the variation of $K_f(T)$ and $\int K_f(T)dT$ with T for unirradiated stoichiometric $(U_{0.8}Pu_{0.2})O_2$ from 500 °C to the melting point. The integral from a fuel surface temperature T_{fs} of 1000 °C up to the melting point is about 4.9 kWm^{-1} for stoichiometric oxide.

Equation 2.3 shows that $\int K_f(T)dT \approx 4.9$ kWm^{-1} implies $q \sim 62$ kWm^{-1} if the fuel has no central hole. But if there is a hole with radius 20% of the fuel radius ($\alpha = .04$) equation 2.5 shows that q can be increased to about 71 kWm^{-1}. In practice, because of the uncertainty of the effects of cracking, porosity, stoichiometry, changes in composition with burnup and the conductance between fuel and cladding (section 2.2.3), q is limited to about 50 kWm^{-1}.

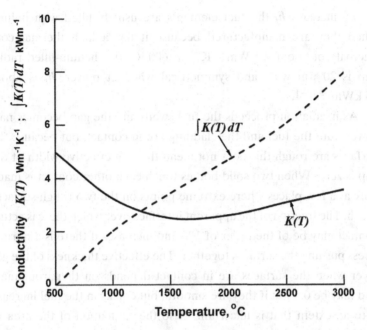

Figure 2.1 The thermal conductivity of $(U_{0.8}Pu_{0.2})O_2$.

2.2.3 Conductance between Fuel and Cladding

A major uncertainty in predicting the fuel temperature is the value of the conductance across the interface or gap between the fuel and the cladding. This is h_f in equation 2.1.

If the fuel is manufactured in the form of pellets they are inserted into the cladding tubes with a radial clearance of some 50 μm or more when cold. This is reduced by the differential thermal expansion when the reactor is at power. The thermal expansion coefficient of the cladding (about $18 \times 10^{-6}\ K^{-1}$) is greater than that of the fuel (about $13 \times 10^{-6}\ K^{-1}$), but as the increase in the fuel temperature is greater the gap is reduced. The extent of the reduction is hard to predict, however, as the expansion of the fuel is irregular because it cracks (see section 2.4.2).

To increase h_f the fuel elements are usually filled with helium when they are manufactured because it has a high thermal conductivity of about 0.3 $Wm^{-1}\,K^{-1}$ at 600 °C. If a helium-filled radial gap is 20 μm wide and symmetrical when at power, h_f is about 15 $kWm^{-2}\,K^{-1}$.

As irradiation proceeds the fuel swells and the gap becomes narrower until the fuel and the cladding are in contact, but because the surfaces are rough this does not mean that the effective width of the gap is zero. When two solid bodies touch each other contact is made only at a few places where extreme points on the two rough surfaces touch. The fraction of the apparent interface over which there is actual contact may be of the order of 1% and increases if there is a normal stress pushing the surfaces together. The effective thickness of the gas layer when the surfaces are in contact depends on their roughness and may be 0.5 μm if they are smooth, but cracks in the fuel increase it to an extent that is uncertain. Also the estimation of the area of solid-to-solid contact is uncertain.

A further large uncertainty is introduced by the changes in the composition of the gas in the gap. As irradiation proceeds inert fission-product gases, mainly xenon with some krypton, are released from the fuel (see section 2.3.5). Because of their high atomic weights they have much lower thermal conductivities than helium (.026 $Wm^{-1}\,K^{-1}$ for krypton and .016 $Wm^{-1}\,K^{-1}$ for xenon at 600 °C), and as they are released from the fuel they reduce the conductivity of the helium substantially by diluting it, introducing another uncertainty in the conductance.

In principle it is possible to estimate h_f experimentally, either by examining fuel after irradiation and deducing the temperature distribution from the observed changes in the structure, or by measuring the fuel temperature during operation by inserting thermocouples into it. Either method involves large uncertainties however, mainly because the conductivity of the fuel itself is not known accurately as explained earlier.

As a result it is usually best to assume an approximate value of h_f of about 5 or 8 kWm^{-2} K^{-1}, and to recognise that it is very uncertain and that the actual value may differ by a factor of two either way. For $R_f = 3$ mm and $q = 50$ kWm^{-1}, $h_f = 6$ kWm^{-2} K^{-1} gives an interfacial temperature difference between fuel and cladding of 440 K, but the actual value may be anywhere between 200 and 800 K.

2.3 DESIGN AND MANUFACTURE OF OXIDE FUEL

2.3.1 Porosity, Swelling and Smear Density

Almost every fission event turns one atom (of uranium or plutonium) into two (of fission products). Some of these are gaseous or volatile (see section 2.3.5) and are mostly lost from the fuel, but the majority are solid and are retained either as interstitial atoms in the crystals or as inclusions of a distinct solid phase. Some of the gaseous fission products are retained as small bubbles within the crystals or on the grain boundaries.

Very roughly all atoms occupy about the same volume (of the order of 10^{-29} m^3) in solid and liquid phases so the increase in the number of atoms means that the volume increases. Loss of some fission products from the fuel tends to mitigate the increase, but on the other hand retention of bubbles of gaseous fission products tends to enhance it. The net result is a fractional increase in the volume of the solid fuel of about 0.8 times the burnup.

This swelling is almost inexorable and if the fuel fits closely inside the cladding when it is new the cladding is forced to strain to accommodate it. Thus for example 10% burnup would imply about 3% normal strain in the cladding in both the circumferential and (because friction between fuel and cladding does not allow relative motion) axial directions. As pointed out in section 3.3.4 the ductility of the irradiated cladding may be too low to accommodate so large a strain. Since the primary purpose of the cladding is to retain the radioactive fission products, if more than a very few fuel elements crack because of the swelling of the fuel inside them the design is unacceptable. The burnup has to be limited by design to a value at which trial irradiation has shown that no more than an acceptable number of cladding failures will occur.

It is usual to allow for the effects of fuel swelling by leaving space for the fuel to swell without straining the cladding. In some cases this has been done by manufacturing the fuel in the form of pellets with a central hole so that the increase in volume can be accommodated by filling or partially filling the hole. Alternatively and more usually the fuel is manufactured as porous pellets, the pores occupying some 10–20% of the overall volume, and the swelling is accommodated by filling the pores.

It is by no means certain, however, that either pores or a central void work in the manner intended to accommodate swelling. As explained in section 2.4.1 pores migrate during operation and a central void is often formed even when one is not present initially. In addition the fuel creeps (see section 2.4.3) so that there is a tendency for stress between fuel and cladding to be relieved by strain of the fuel as well as of the cladding, and irradiation-induced reduction of the density of the cladding (section 3.3.2) also helps to accommodate fuel swelling. In fact the picture is so confused that, although the strain of the cladding of an irradiated fuel element can be measured, it is not easy to predict.

A convenient parameter often used to characterise the space available to accommodate the swelling of the fuel is the "smear density". It is often defined as the ratio between the mass of fuel per unit length of the fuel element as manufactured and the mass of fuel it would contain if the cladding were completely filled with fuel at its maximum theoretical density (i.e. if there were no gap between the pellets and the cladding, and no porosity or central hole in the pellets). Oxide fuel elements are typically manufactured with 80% smear density.

Sometimes the term is used slightly differently, being defined as the average density the fuel would have if it were spread uniformly across the cross-section of the fuel element. Thus if the theoretical density of UO_2 is 11000 kgm^{-3}, a fuel element with 80% smear density according to the definition in the previous paragraph might alternatively be said to have a smear density of 8800 kgm^{-3}.

2.3.2 Manufacturing Processes

In most cases the fuel is in the form of sintered pellets that are made from a mixture of UO_2 and PuO_2 powders in the correct proportion. It is important that the powders are mixed homogeneously because if they are not there is a risk of hot spots due to high local plutonium concentration. There is also the danger that if the power increases very rapidly in an accident the effect of the negative Doppler coefficient of reactivity from the ^{238}U (see section 1.6.5) may be reduced if there is a significant delay in transferring heat from the plutonium-rich regions, where most of the fissions occur.

The oxide power is mixed with a binder or plasticiser, usually an organic compound, which allows the powder to be pressed into the required shape. The pellets are then sintered to turn them into the required ceramic form. During the sintering process the binder is broken down and driven off, and care has to be taken to remove any traces of hydrogen and carbon, which can have deleterious effects on the cladding. The porosity of the finished pellets can be controlled by careful choice of the nature and quantity of the binder and of the sintering time and temperature, and it can be made as low as 2% if required. Finally the pellets may have to be ground to the correct size. This is expensive because it produces plutonium-bearing dust which is hazardous and therefore has to be controlled carefully. Precise control of the sintering process to produce pellets of the required dimensions within a sufficiently close tolerance is preferable.

The oxygen content of the fuel can be controlled at the sintering stage. If the sintering is carried out in a reducing atmosphere, containing hydrogen for example, the oxygen can be reduced to a sub-stoichiometric level (i.e. $x < 0$).

An alternative manufacturing technique uses the "sol-gel" method, in which plutonium and uranium nitrate solutions are mixed in the correct proportions. This aqueous mixture is sprayed through a vibrating nozzle to produce small gel droplets that are then dried to produce oxide microspheres in which the plutonium is in solid solution within the $(U,Pu)O_2$ crystals. These microspheres can then be sintered into pellets.

Instead of being made into pellets the sol-gel microspheres can be used to make "vibro-compacted" or "vipac" fuel. In this technique microspheres of at least two different sizes are loaded directly into the cladding tubes, which are vibrated until they are compacted to the required density and then sintered. A disadvantage of vipac fuel is that even if the porosity of the individual particles is as low as 3% the overall porosity of a 6-mm diameter fuel element made with 800 and 80 μm microspheres cannot be reduced below about 18%. To achieve a higher smear density a third, even smaller, particle size is required, which increases the cost of manufacture significantly. Another disadvantage is that if the small particles, or "fines" as they are called, contain plutonium they are hazardous in manufacture because of the risk of inhalation. If the plutonium is confined to the larger particles the hazard is reduced but the disadvantages of possible hot-spots and a delay in the Doppler effect are incurred.

2.3.3 Reprocessing

Design and manufacture of the fuel have to take account of the way it is to be reprocessed after discharge from the reactor (assuming it is not to be consigned to long-term storage as waste material). Oxide fuel is usually reprocessed by the "purex" process, which involves dissolving it in nitric acid. However $(U,Pu)O_2$ containing more than about 40% plutonium does not dissolve readily. It is therefore necessary to ensure that the fuel is mixed uniformly during manufacture, because particles of fuel with a high plutonium content might not dissolve and would cause difficulty at some later stage in the process. It is not difficult to ensure uniformity of co-precipitated fuel but if pellets are formed from mixed UO_2 and PuO_2 powders care has to be taken that the grains are very small and are mixed thoroughly. In vipac fuel all the plutonium may be in the larger granules so their plutonium concentration has to be greater than the average for the fuel as a whole, and if after irradiation (when the plutonium concentration is still higher) it exceeds 40% difficulty will be experienced in dissolution.

Purex is a solvent extraction process that separates first uranium and plutonium from the fission products, and then uranium and plutonium from each other. The separation depends on differences between solubility in water and in an organic solvent, and these differences depend on acidity. The feed for the process is the aqueous stream of fuel dissolved in nitric acid, containing the uranium, plutonium and fission products as nitrates, which is brought into contact with a stream of the organic solvent tributyl phosphate (TBP). TBP and water are not mutually soluble. Under acid conditions uranium and plutonium are more soluble in TBP whereas the fission products are more soluble in water, so when the aqueous and organic phases are agitated together the heavy metals are transferred to the TBP while the fission products remain in the water. The two phases are then separated and the TBP stream is brought into contact with water of neutral acidity where the heavy metals go back into the water. The process can be repeated in a second cycle to remove any residual fission products. With two cycles a decontamination factor (the ratio of the fission-product concentration in the final heavy metal product to that in the initial feed) of 10^{-6} can be achieved.

Plutonium can be separated from uranium by making use of the different valency states it can take. Tri-valent plutonium in the form of $Pu(NO_3)_3$ is not soluble in TBP, so by adding a reagent that reduces Pu(IV) to Pu(III), uranium can be taken into the organic phase, leaving plutonium in the aqueous. Separation factors of 10^{-4} can be achieved.

These solvent extraction processes take place in contactors that agitate the immiscible aqueous and organic streams together in such a way as to make the area of the interface between them as large as possible. The contactors may be horizontal "mixer-settler" tanks fitted with alternating agitated and quiescent compartments, but more usually are vertical packed columns. In a "pulsed column" the effect of the packing or perforated plates in the column as they break up the streams of TBP (flowing upwards) and water (flowing down) is enhanced by pulsing the feed flow.

Whatever the form of the contactors they have to be designed to avoid criticality. This can be done by means of the geometry (for example by minimising the diameter of a pulsed column) and by incorporating structural materials that contain thermal neutron absorbers such as boron or gadolinium.

2.3.4 Stoichiometry and Oxygen Potential

Corrosion of the cladding by the fuel is discussed in section 2.4.7. It is strongly affected by the oxygen potential in the fuel and therefore by the stoichiometry. Figure 2.2 shows the variation of the oxygen potential of $(U_{1-a}Pu_a)O_{2+x}$ with x at constant temperature for various values of a. As expected the oxygen potential rises with x, but also it is higher for PuO_2 ($a = 1$) than for UO_2 ($a = 0$) because uranium adopts higher valency states than plutonium. The steep rise in the oxygen potential of UO_{2+x} as x increases through zero is important.

The variation of oxygen potential with temperature is shown in Figure 2.3. Oxygen tends to migrate to the cooler parts of the fuel if $x < 0$, but if $x > 0$ the tendency is considerably reduced.

The effect of burnup is very complicated because of the wide range of elements formed as fission products. Each fission releases two

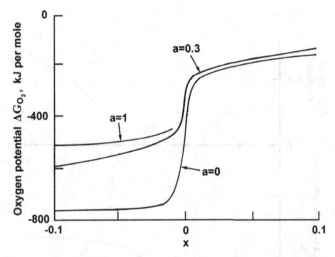

Figure 2.2 The oxygen potential of $(U_{1-a}Pu_a)O_2$ at 1500 °C.

oxygen atoms, some of which go to oxidise the fission products, such as zirconium, strontium, barium and the rare earths, which have a strong affinity for oxygen. The number of oxygen atoms taken up by this oxidisation process depends on the yields of the various fission

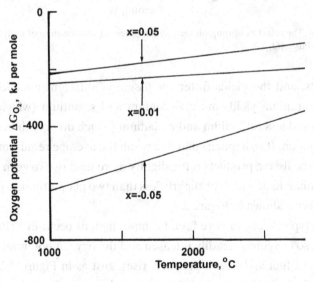

Figure 2.3 The effect of temperature on the oxygen potential of $(U_{0.7}Pu_{0.3})O_2$.

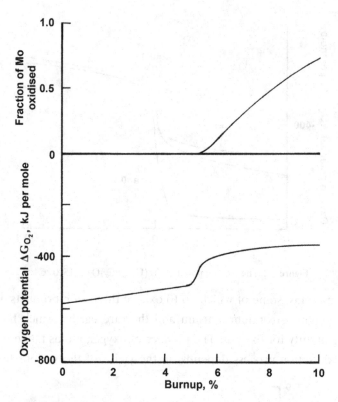

Figure 2.4 The effect of burnup on the oxygen potential and the state of molybdenum for $(U_{0.85}Pu_{0.15})O_{1.96}$.

products, and the yields differ for fission of different nuclides. Fission of uranium yields more zirconium and strontium (which form oxides) and less ruthenium and palladium (which do not) than fission of plutonium. It so happens that as a result the average requirement to oxidise the fission products is for slightly more than two oxygen atoms per uranium fission and for slightly less than two per plutonium fission. The effect is shown in Figure 2.4.

In a typical reactor core fuel, far more fissions occur in ^{239}Pu than in ^{238}U, so oxygen is steadily released and the oxygen potential of the mixture of fuel and fission products rises. Just as in Figure 2.2 there is a particularly sharp rise at the point in burnup where the oxygen

Table 2.1 *Fission yields of long-lived isotopes of inert gases*

Isotope	Half-life	Fission yield (atoms per fission)
^{83}Kr	Stable	.005
^{84}Kr	"	.010
^{85}Kr	10.3 years	.003
^{86}Kr	Stable	.020
^{131}Xe	Stable	.029
^{132}Xe	"	.043
^{134}Xe	"	.080
^{136}Xe	"	.064

content becomes stoichiometric, but the oxygen potential does not rise very high. An effective upper limit is set by molybdenum. The Mo/MoO_2 system has an oxygen potential very similar to that of the fuel, and because the yield of molybdenum is high (about 0.06 atoms for each fast fission of either ^{239}Pu or ^{238}U) it forms a buffer. When the stoichiometric ratio is reached, oxygen released by further fissions is taken up by oxidising molybdenum.

To minimise oxidative corrosion of the cladding the fuel is usually manufactured a few percent sub-stoichiometric, with $x = -0.02$ or -0.03, but as explained in section 2.4.7 this does not prevent some corrosion taking place.

2.3.5 Fission-Product Gas Release

Various isotopes of krypton and xenon are produced by fission. The cumulative yields of the stable and long-lived isotopes are shown in Table 2.1. Most of them are formed not immediately but as the end-products of beta-decay chains. The half-lives of most of the decay processes are of the order of a few minutes, and in only two cases

(^{131}Xe and ^{132}Xe) of a few days. The total yield is about 0.04 atoms of krypton and 0.22 atoms of xenon per fission.

This is a very large quantity. If fuel of density 10^4 kg m^{-3} undergoes 10% burnup and the krypton and xenon generated exist as gas at 400 °C and atmospheric pressure they occupy 53 times the volume of the fuel. Alternatively if they are confined in a volume equal to that of the fuel, they exert a pressure of 5.3 MPa at 400 °C or 6.8 MPa at 600 °C.

The krypton and xenon are not formed as gases, however, but as single atoms in solution in the fuel. The solubility is very low so the solution is soon supersaturated and the atoms are precipitated to form bubbles. They enter the bubbles by thermal diffusion (and therefore at a rate that depends on temperature), and they may also leave the bubbles and return to the fuel by the process known as fission-induced resolution. This comes about as follows. If a fission occurs close to a gas bubble one of the fission-fragments may enter it. It may have 100 MeV or more of kinetic energy, and if it collides with a gas atom it can easily impart enough energy to it to make it reenter the surrounding fuel crystals. In effect the gas atom is knocked out of the bubble and forced back into solution.

The size and rate of growth of a bubble depend on the rates at which it gains and loses gas atoms. Bubbles within a grain gain by diffusion and also by sweeping up atoms or other bubbles as they move through the grain (see section 2.4.1 for the mechanism of this process) and lose by resolution. They are typically spherical with a diameter of the order of 5 nm. Bubbles at the grain boundaries, however, grow as the grain boundaries move. The hotter part of the fuel is continually being recrystallised, and as one grain grows at the expense of a neighbour the grain boundary sweeps through the fuel collecting gas atoms and bubbles as it goes. This is a very effective growth mechanism and as a result grain-boundary bubbles are larger than bubbles within the grains. Grain-boundary bubbles are sometimes flat or lenticular in shape and up to 20 nm across. Figure 2.5 shows typical gas bubbles. As the grain-boundary bubbles grow they link up and eventually form

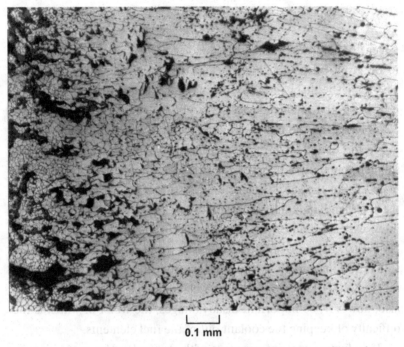

0.1 mm

Figure 2.5 Gas bubbles in irradiated oxide fuel (0.8% burnup at 50 kWm^{-1}).

channels by which the gas escapes from the fuel. The fraction of the gas that is released in this way depends in a complex manner on burnup and temperature. Because the process is so complicated it is usual to assume when designing the fuel elements that all the krypton and xenon generated will be released as gas.

In reality, however, some 10 or 20% is retained in the fuel, and this may become very important in the course of an accident. If the fuel becomes overheated gas retained in the colder parts may suddenly be released and affect the subsequent course of events (see section 5.4.5). An approximate estimate of the amount available can be made by assuming that 2% of the gas generated in fuel at temperatures above 1800 °C is retained, 50% in fuel between 1800 °C and 1400 °C, and 70% in fuel below 1400 °C.

2.3.6 Sealed or Vented Fuel

Given that allowance has to be made for this large quantity of gas to be given off by the fuel during irradiation the next question is whether it should be retained within the fuel element or released from it. If it is to be retained space has to be provided for it. If a volume equal to that of the fuel is provided, after 10% burnup the pressure inside the fuel element may be some 5–7 MPa as we have seen. A smaller volume would necessitate thicker cladding; a larger volume would make the fuel element longer so that the reactor vessel and the fuel handling equipment would have to be larger and therefore more expensive and the pressure drop in the coolant circuit would be greater. If the gas could be released from the fuel elements there would be considerable advantages in a reduction of the height of the reactor vessel and possibly of the cladding thickness. There would also, however, be important disadvantages, the worst of which is the difficulty of keeping the coolant out of the fuel elements.

If sodium comes into contact with the fuel under some circumstances a mixture of sodium uranate and plutonate (Na_3UO_4 and Na_3PuO_4) is formed. These compounds have low densities causing the fuel to swell and possibly bursting the cladding. If the gas is to be released, therefore, each fuel element has to be fitted with a vent to allow the gas to leave while preventing coolant from entering even when the reactor is shut down and cooled and the pressure inside the fuel elements falls. It has not proved possible to design a vent that is both sufficiently reliable and sufficiently cheap to manufacture in large numbers.

A second disadvantage of vented fuel is that it inevitably allows the release of radioactive materials into the coolant. In addition to the isotopes listed in Table 2.1 various short-lived isotopes of krypton and xenon such as ^{90}Kr (half-life 33 s), metastable ^{83}Kr (114 min) and ^{133}Xe (5.3 d) appear as fission products. To prevent or minimise release of these isotopes a device would have to be incorporated to delay the release of the gases until they have decayed, because the presence in the coolant of their decay products, such as ^{90}Sr of which ^{90}Kr is a precursor, is very undesirable. Of course nothing could be done to prevent the release of ^{85}Kr from vented fuel and it would be very hard to guarantee that other volatile fission products such as ^{131}I (8 d) and ^{137}Cs (29 yr), of which the yields are high, would not be released with the inert gases.

2.3.7 Fuel Element Design

An important choice facing the designer of sealed fuel elements is whether to place the gas storage volume, or plenum as it is usually called, above or below the core. If it is below, surrounded by cooler inlet coolant, it can be smaller than if it is above, where it is immersed in the hotter outlet coolant. But if for some reason the plenum should burst or leak the gas would be released and pass upwards, displacing coolant from the core and possibly causing a positive reactivity transient (see Figure 1.26).

For a breeder reactor there is usually an axial breeder region containing natural or depleted UO_2 between the core fuel and the plenum. Provision has to be made to allow the gas from the core to pass through the breeder to the plenum. The axial breeder on the other side of the core (i.e. above the core if the plenum is below) can be incorporated in the same fuel elements with the core. Alternatively it can be made in the form of separate fuel elements that, because of the low power density in the breeder, can have a larger diameter than the core elements. The advantage of the latter arrangement is that the resistance to coolant flow is reduced, but it increases the complexity.

Figure 2.6 shows a typical core fuel element that incorporates both axial breeders, has the fuel in the form of pellets, and has the plenum below the core. The retainers are necessary only to keep the fuel pellets in position during manufacture, transport and loading into the reactor.

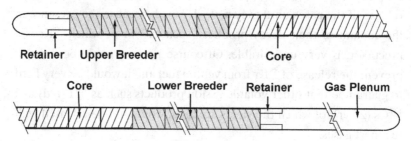

Figure 2.6 A typical fuel element.

Almost as soon as the power is raised the pellets swell to press against the cladding (see section 2.4.4) and become jammed.

The diameter of the fuel element is determined by heat transfer and manufacturing cost considerations (section 3.2.1). The thickness of the cladding has to be sufficient after allowing for corrosion both on the outside (sections 3.3.4 and 3.3.5) and the inside (section 2.4.7) to withstand the stresses due to fuel swelling and fission-product gas. Its thickness is usually about 0.3 or 0.4 mm.

Radial breeder fuel elements are usually similar in form to those of the core but with larger diameter. Even at the end of several years of irradiation the power density in the breeder adjacent to the core is only 20% or so of that at the core centre. The breeder elements can therefore have more than twice the diameter of the core elements before the limiting linear heat rating is reached.

2.4 IRRADIATION BEHAVIOUR OF OXIDE FUEL

2.4.1 Recrystallisation

During irradiation the crystal structure of the fuel is changed almost completely. Figure 2.7 shows a polished and etched cross-section of typical pellet fuel after irradiation and Figure 2.8 shows a similar axial cross-section of vipac fuel. Throughout most of the cross-section the

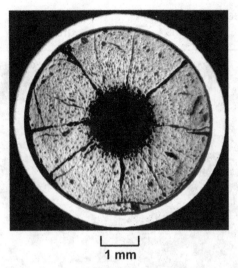

Figure 2.7 Cross-section of a fuel element irradiated to 7.2% burnup at $40\,\mathrm{kWm^{-1}}$.

original particulate structure of the vipac fuel has completely disappeared.

The most striking change is that a hole appears in the centre. It is surrounded by a region of high density in which the grains are

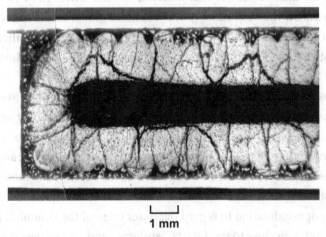

Figure 2.8 Axial section of a vipac fuel element irraditated to 4.6% burnup at $31\,\mathrm{kWm^{-1}}$.

Unrestructured   Equiaxial   Columnar
grains   grains   grains

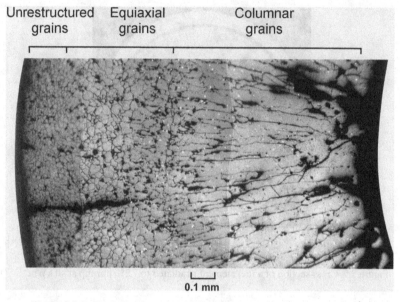

0.1 mm

Figure 2.9 Recrystallisation of fuel irradiated to 8% burnup at 40 kWm^{-1}.

long and narrow and lie along radii of the cylinder. This is known as the "columnar grain" region. Outside it is a region where the grains are larger than in the original material but are oriented at random. This is called the "equiaxial grain" region. Finally, in the outermost part of the fuel it retains its original structure, and this is called the "unrestructured" region. These three distinct regions are shown in Figure 2.9.

Observations made after various periods of irradiation show that this basic structure is set up very quickly – within an hour of achieving the full power density. Thereafter the pattern of equiaxial and columnar grains spreads outwards from the centre but at a rate that decreases very rapidly. The higher the linear rating the greater the radii of the restructured regions, and for most of the life of the fuel it is a fair approximation to regard the outer edge of the columnar grain region as being close to the 1800 °C isotherm, and that of the equiaxial region close to the 1600 °C isotherm.

As far as is known the restructuring takes place as follows. At intermediate temperatures the sintering that started during manufacture continues. The pores between the grains tend to coalesce and the individual grains grow to form the equiaxial grain region.

At higher temperatures the pores become mobile and move up the temperature gradient. There are two mechanisms for this process: surface diffusion whereby atoms of fuel move round the pore from the hot to the cold side, and volume diffusion whereby atoms evaporate from the hot side and condense on the cold. The result is that the pore moves towards the region of higher temperature and at the same time the fuel is recrystallised as the pore moves through it. The atoms deposited on the cold side of the pore form a single new crystal, relatively free from imperfections and occupying the volume swept out by the pore as it moves.

The pores move to the centre – the hottest part – of the fuel and there form the central void, while the density of the columnar grain region increases to something close to the theoretical value, which is that of a single crystal. The speed with which the pores move depends strongly on temperature so the outer boundary of the columnar region moves quickly to start with and then very slowly. Initially it is the pores incorporated in the material when it is manufactured that move in this way but later it is the bubbles formed by the accumulation of fission-product gas on the grain boundaries (section 2.3.1). As a result the columnar grain region is continually being recrystallised.

2.4.2 Cracking

The temperature difference of 1500 K or more between centre and surface of the fuel induces substantial thermal stresses that cause cracks. Figure 2.7 shows a cracking pattern typical of those seen in irradiated fuel. It is important to realise, however, that the pattern of cracks seen when the fuel is cool is quite different from that present during irradiation. The conjectured development of cracks in pellet fuel is shown

(a) Startup   (b) After "healing"   (c) Shutdown

Figure 2.10 The development of cracks in fuel during irradiation.

in Figure 2.10. When the fuel first generates power and experiences a temperature difference stresses are set up. The tensile hoop stress at the outside exceeds the rupture stress and radial cracks are formed as indicated in Figure 2.10a. In the centre, however, cracks are very quickly healed by the recrystallisation process.

After a certain amount of irradiation, usually about 0.1% burnup, the fuel has swollen to touch the cladding. Subsequent swelling is resisted by the cladding and the fuel is subject to a compressive stress. Under the influence of this it creeps rapidly near the centre and very slowly at the periphery. As a result the cracks close at the centre but not at the outside. When they have closed sintering or recrystallisation heals them. Thus after prolonged irradiation the centre of the fuel is free of cracks while the periphery retains cracks that tend to taper inwards (Figure 2.10b).

When the reactor is shut down the fuel cools, the centre contracts more than the periphery, and new radial cracks are formed that are wide at the centre and taper towards the outside (Figure 2.10c). These cracks are prominent in pictures of fuel cross-sections, such as Figure 2.7, but it should be remembered that they were not present when the reactor was operating.

2.4.3 Thermal and Irradiation Creep

Thermal creep in the fuel is usually described by an empirical relationship of the form

$$\dot{\varepsilon}_{th} = A\sigma^n e^{-B/T}, \tag{2.6}$$

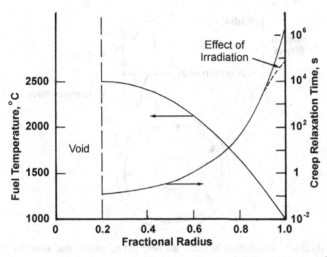

Figure 2.11 Creep relaxation time in a fuel element generating 45 kWm^{-1}.

where σ is the stress, ε is the strain, T is temperature and A, B and n are constants. Usually $n = 1$, B is about 4×10^4 K, and typical results for stoichiometric UO_2 ($x = 0$) give $A \approx 2 \times 10^{-4}$ s^{-1} Pa^{-1}. If the fuel is not stoichiometric it is harder, whereas mixed oxide is usually found to be softer.

The resulting relaxation time for thermal stress is shown in Figure 2.11. It is very short in the centre of the fuel, and over times of the order of seconds or more this part of the fuel can support only hydrostatic pressure. The outer part of the fuel on the other hand is quite rigid.

A second source of creep strain is due to the irradiation itself. The effect of a fission event is to melt the small volume of the fuel through which the fission fragments travel. Any shear stress in the melted region is transferred to the surrounding material causing it to strain a little. Then another fission event melts another region causing another additional strain, and so on. The result is a steady strain-rate proportional to stress and fission-rate density but only weakly dependent on temperature. Various experimental correlations have been proposed, typically of the form

$$\dot{\varepsilon}_f = A_1 f \sigma \tag{2.7}$$

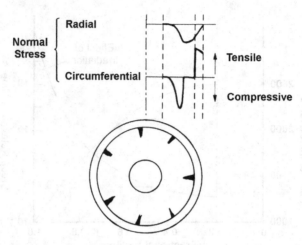

Figure 2.12 Stress distribution in a fuel element during irradiation, after the fuel has made contact with the cladding.

with $A_1 \approx 1.8 \times 10^{-36}$ m^3 Pa^{-1} · f is the fission-rate density in m^{-3} s^{-1}, so with σ in Pa equation 2.7 gives $\dot{\varepsilon}$ in s^{-1}.

The effect of irradiation creep is also shown in Figure 2.11. The relaxation time in the outer part of the fuel is substantially reduced but it is still hard on a timescale of days or weeks.

It is of course impossible to observe what happens to the fuel during irradiation but it is probably more or less as follows. After the fuel has swollen to touch the cladding the cladding exerts a compressive normal stress on its surface. The resulting stress distribution is shown in Figure 2.12. The fuel is incapable of exerting hoop stress in the central region, which is soft and stress-free because it is in contact with the central void. It is also incapable of exerting hoop stress at the outer edge where there are radial cracks. Thus the compressive surface stress is borne by a compressive hoop stress in a narrow ring of fuel just at the root of the cracks.

In steady operation the load-bearing ring moves slowly outwards as more of the fuel has time to relax the imposed stress and allow the cracks to close. But if there are changes in reactor power and therefore in the fuel temperature distribution the development is interrupted.

Figure 2.13 Distortion of a fuel pellet during irradiation (exaggerated).

If the power rises, for example, the cracks reopen to some extent and the load-bearing ring is moved inwards.

2.4.4 Interaction between Fuel and Cladding

The fuel swells at a roughly constant rate throughout irradiation and has usually closed the gap and come into contact with the cladding after about 0.1% burnup. The cladding also swells (section 3.3.2) but only slowly at first, more rapidly later, and at a rate that depends strongly on temperature. The result is that for some cladding materials and in some parts of the core the swelling rate of the cladding may eventually exceed that of the fuel, tending to reduce the stress between the two and reopen the gap.

While cladding and fuel are in contact and there is a compressive stress between them as shown in Figure 2.12 both creep. Swelling of the central part of the fuel inside the load-bearing ring is accommodated by expansion into the central void. Swelling of the outer part is accommodated partly by the swelling of the cladding.

The cladding strain is not always uniform and in some cases may be concentrated by cracks in the fuel pellets, and this stress concentration, rather than uniform strain, may determine the maximum burnup to which the fuel can be subjected. Sometimes distortion of the pellets can cause non-uniform strain of the cladding. During irradiation a pellet tends to distort into the shape shown, very much exaggerated, in Figure 2.13. The ends tend to be displaced outwards as shown and can

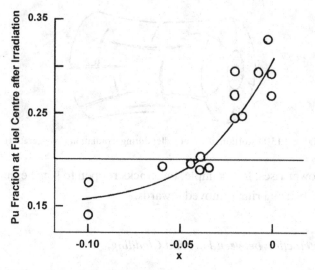

Figure 2.14 The effect of the initial ratio of oxygen to metal on the migration of plutonium in fuel with initial composition $(U_{0.8}Pu_{0.2})O_{2+x}$.

cause ridges round the cladding, giving it the appearance of a bamboo cane. These ridges are not usually permanent because the pellets soon fuse together, but they are sometimes observed in fuel that has been subjected to abnormal conditions.

2.4.5 Migration of Plutonium and Oxygen

It is found that, even if the plutonium is initially distributed uniformly throughout the fuel, after only a short irradiation (less than 1% burnup) the relative concentrations of uranium and plutonium have changed, but the nature of the redistribution depends on x. Figure 2.14 shows experimental values of plutonium concentration at the centre of the fuel after irradiation, indicating that for $x < -0.04$ the plutonium moves outwards, whereas for $x > -0.04$ it moves inwards. In contrast the plutonium in the outer, cooler, part of the fuel is hardly affected. The distribution of plutonium in irradiated fuel with $x = 0$ is shown in Figure 2.15.

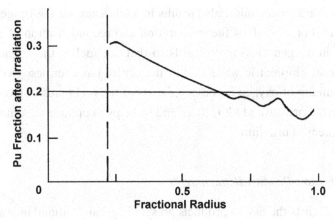

Figure 2.15 The redistribution of plutonium in fuel with initial composition $(U_{0.8}Pu_{0.2})O_{2+x}$.

The mechanism for this redistribution seems to depend on the preferential evaporation and migration of different species. The stoichiometric ratio is important because if there is excess oxygen the mixture of vapours in equilibrium with the mixed oxide contains relatively more UO_3, whereas if oxygen is deficient it contains more PuO, and for $x < -0.03$ or -0.04 PuO predominates. Thus for more positive values of x it is uranium in the form of UO_3 that tends to evaporate from the hotter part of the fuel and condense in the cooler, whereas for more negative x plutonium in the form of PuO moves instead. This is not the whole story, however, because it does not explain why the segregation is so limited in extent. It may be that diffusion of vapour through the dense columnar grain region is in fact severely restricted.

If the segregation of plutonium and uranium were more severe it would have serious consequences. Because the plutonium generates most of the power its extensive migration to the centre would increase the fuel central temperature and possibly cause melting. It would also have the effect of delaying the Doppler feedback on reactivity (see section 2.3.2). For the limited segregation that is actually observed these effects are negligible.

The same mechanism also results in a tendency for the oxygen to migrate. For $x > -0.04$ the evaporation and recondensation of UO_3 result in oxygen moving outwards so that the fuel at the outside is nearer stoichiometric while that at the centre has even less oxygen. The buildup of oxygen in the outer parts of the fuel tends to increase the partial pressure of UO_3 there and so helps to oppose the outward movement of uranium.

2.4.6 Fission-Product Behaviour

Table 2.2 lists the fission products present in greatest abundance after 10% burnup of typical fast reactor fuel (30% plutonium with typical concentrations of the higher plutonium isotopes). The behaviour of a chemical system with so many components is obviously extremely complex and is certainly not understood in detail. The broad outlines are given in this section but the complexities are such that the actual behaviour in a particular fuel element with a slightly different composition, cladding, temperature or irradiation history may differ quite widely. It is convenient to divide the most abundant fission products into groups as follows. Elements in the same group behave roughly similarly.

   Inert Gases (Kr, Xe). These are mainly released from the fuel, but some are retained in solution or in small bubbles within the grains in the cooler parts of the fuel (see section 2.3.4).

   Alkali Metals (Rb, Cs). These are very volatile in elemental form and migrate to the cool periphery of the fuel. This migration is illustrated in Figure 2.16 which shows the distribution of ^{137}Cs as determined by γ – spectroscopy. In some cases the isotopes that are daughters of inert gases, such as ^{133}Cs which is produced from ^{133}Xe which decays with a half-life of 5.3 d, and ^{87}Rb produced from ^{87}Kr decaying with a half-life of 78 min, can appear in the gas plenum to which the precursors have migrated.

Table 2.2 *Fission-product concentrations*
after 10% burnup

Element		Concentration (atoms per initial heavy atom)
Krypton	Kr	.0020
Rubidium	Rb	.0017
Strontium	Sr	.0039
Yttrium	Y	.0021
Zirconium	Zr	.0202
Molybdenum	Mo	.0206
Technetium	Tc	.0058
Ruthenium	Ru	.0211
Rhodium	Rh	.0052
Palladium	Pd	.0137
Silver	Ag	.0016
Tellurium	Te	.0032
Iodine	I	.0016
Xenon	Xe	.0205
Caesium	Cs	.0187
Barium	Ba	.0065
Lanthanum	La	.0054
Cerium	Ce	.0129
Praseodymium	Pr	.0042
Neodymium	Nd	.0144
Promethium	Pm	.0017
Samarium	Sm	.0037

Note: Only elements with concentrations greater than 10^{-3} are listed.

Halogen (I). It is very difficult to determine what happens to the iodine because stable or long-lived isotopes are produced only in very small quantities (this applies to bromine as well), and ^{131}I decays with a half-life of 8.04 d. Because it is volatile it seems to collect near the cladding and also at the ends of the fuel pins. It may be present as caesium iodide. Iodine may be involved in corrosion of the cladding.

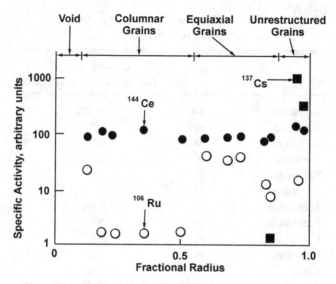

Figure 2.16 The distribution of fission products in irradiated fuel.

Metals forming refractory oxides (Sr, Y, Zr, Ba, La, Ce, Pr, Nd, Pm, Sm). By and large, having formed oxides these elements do not migrate and are found uniformly distributed through the fuel, as illustrated by the data for ^{144}Ce in Figure 2.16. But again the isotopes that are daughters of volatile or gaseous precursors, such as ^{138}Ba and ^{140}Ba (from ^{138}Cs and ^{140}Cs respectively), are less uniform. ^{89}Sr (a daughter of ^{89}Rb that has a half-life of 15.4 min) migrates farther than ^{90}Sr (from ^{90}Rb, half-life 2.7 min).

Metals that do not form oxides (Tc, Ru, Rh, Pd, Ag, Te). These are found as metallic inclusions, sometimes dispersed through the fuel and sometimes, especially if the central temperature is high, having migrated to the central void. There they form droplets of metal that are molten while the reactor is operating and are found as small ingots when the fuel is examined subsequently. Such ingots usually contain uranium and plutonium as well. Figure 2.16 shows the loss of ruthenium from the hottest part of the fuel.

Metal that may form an oxide (Mo). The fate of the molybdenum depends on the oxygen potential of the fuel as explained in section 2.3.3. If it is low molybdenum is found in the metallic inclusions; if it is high it is found as MoO_2.

2.4.7 Corrosion of the Cladding

There are various possible cladding materials but most if not all are steel alloys containing at least a few percent of chromium. After the start of irradiation the inside surface of the cladding is quickly covered with a thin layer of oxide, mainly Cr_2O_3. Further oxidative corrosion involves reactions in the matrix of the metal crystals, often preceded by intergranular attack at the interfaces between crystals.

Fission products are involved particularly in the intergranular attack. The mechanisms are not understood in detail but it is clear that caesium and tellurium, possibly in the form of Cs_2Te, are important. The Cs_2Te appears to react with chromium and the tellurium released then attacks the steel by dissolving iron and nickel.

Intergranular corrosion is facilitated by a high power rating (implying high fuel temperatures) and the presence of an open gap between the fuel and the cladding. High temperature is important because it enhances the mobility of volatile species, allowing the caesium to move out of the fuel matrix to the cladding (see Figure 2.16). The gap between fuel and cladding is open early in the life of the fuel element, but then closes, only to reopen later (see section 2.4.4).

Corrosion at the start of irradiation of a new fuel element, which can occur within a few days, is sometimes known as "_corrosion de jeunesse_". In later life, when the gap has reopened, intergranular corrosion is sometimes seen at the top of the fuel column, where the surface temperature of the fuel is highest, and is sometimes called "_reaction interface fissile-fertile_", or RIFF.

In spite of uncertainty about the details of the mechanisms of corrosion it is clear that corrosion increases with burnup, temperature gradients, and oxygen content. The last is most important and is the principal reason for making the oxygen content of new fuel significantly sub-stoichiometric, x being typically -0.02 or -0.03. And in spite of the propensity for intergranular corrosion experience has shown that it rarely leads to loss of integrity of the cladding, even though it is stressed by the pressure of the fission-product gases. Many tens of thousands of oxide fuel elements have been irradiated to burnups as high as 25% or more, with failure rates of the order of 0.1% or less.

2.5 METAL FUEL

2.5.1 Temperatures

In the early days of the development of fast reactors, when a high breeding ratio was thought to be of great importance, it was realised that the best fuel material from the point of view of the neutron economy would be a metal, either uranium or an alloy of uranium and plutonium. By the 1960s however it was realised that high burnup was important, and as at that time metal fuel was thought to be limited to about 3% burnup interest almost everywhere turned to oxide. The sole exception was in the USA where work continued in support of the metal-fuelled EBR-II test reactor. That work was ultimately successful in that a ternary U-Pu-Zr alloy fuel capable of up to 20% burnup was developed, and is now an alternative to oxide.

Figure 2.17 shows the thermal conductivity of solid 70U-20Pu-10Zr (w/o), and its integral, between 500 and 900 °C. There are few data for higher temperatures. However as explained in section 2.5.2 on irradiation the metal becomes very porous, and a porosity of 30% may reduce the effective conductivity to 40% of the value for the solid material. A value of 10 $Wm^{-1} K^{-1}$ is a reasonable conservative approximation.

The melting points of pure uranium and plutonium metals are 1135 °C and 640 °C respectively. The *solidus* temperatures of 20%

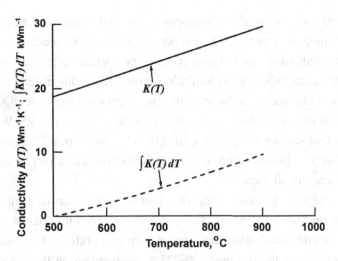

Figure 2.17 The thermal conductivity of 70U-20Pu-10Zr alloy.

and 40% Pu in U are 920 °C and 780 °C respectively and the eutectic (85%Pu) is about 620 °C. The admixture of zirconium increases the *solidus* temperature of U-20Pu by about 13 K for each percent. Thus the *solidus* temperature for 70U-20Pu-10Zr is about 1050 °C.

Unlike oxide (see section 2.3.6) metal fuel is chemically compatible with sodium, so the thermal conductance between fuel and cladding can be improved by filling the gap with sodium. As a result the temperature difference between the coolant and the outer surface of the fuel is very small. For a linear rating of 50 kWm^{-1}, steel cladding 0.3 mm thick and a 0.35 mm sodium-filled gap between fuel and cladding, the difference is only 53 K. At the centre of a reactor core where the power density is greatest and the coolant temperature is 500 °C the fuel surface temperature might therefore be about 550 °C. If melting is to be avoided the temperature difference between the surface and the centre of the fuel cannot be allowed to exceed 1050–550 = 500 K. Using a conservative conductivity of 10 Wm^{-1} K^{-1} the linear heat rating $q = 4\pi K_f (\Delta T_f)$ has to be limited to ~63 kWm^{-1}. In practice q is limited to about 50 kWm^{-1} to provide a margin to melting and to allow for uncertainty over the effect of porosity on the conductivity.

Thus in spite of the differences in thermal properties the linear heat rating in a metal-fuelled reactor is likely to be similar to that in one with oxide fuel. There is however a great difference in the fuel temperatures. The volumetric average fuel temperature at the centre of the reactor of the previous paragraph would be about 800 °C, whereas that for a similar oxide-fuelled reactor (with $q = 50$ kWm^{-1} and a fuel surface temperature of 1000 °C – see section 2.2.2) would be 1700 °C. This difference has a great effect on the behaviour of the fuel under irradiation.

There may be come concern about the formation of a eutectic where the fuel touches the cladding, and the possibility that the resulting liquid metal might damage or even penetrate it. The eutectic temperature is in the range 700–725 °C (depending on the composition of the fuel). Even if the coolant temperature at the core outlet is 600 °C the maximum fuel surface temperature in an element with a peak linear rating of 50 kWm^{-1} (i.e. about 25 kWm^{-1} at the top of the core) would be around 625 °C, leaving a substantial margin to eutectic formation.

2.5.2 Swelling

Uranium metal adopts different crystalline forms at different temperatures. The stable phases are an α-phase consisting of anisotropic orthorhombic crystals below 668 °C, a tetragonal β-phase from 668 °C to 776 °C, and a γ-phase consisting of isotropic cubic crystals from 776 °C up to the melting point. Plutonium is soluble in concentrations up to 16% in the α-phase and completely soluble in the γ-phase.

The α-phase is dimensionally unstable under irradiation. The high-energy fission fragments from each fission event displace some of the metal atoms from their positions in the crystal lattice as they slow down, in effect melting a small volume of the crystal. As this cools and resolidifies the shape of the crystal changes and it grows anisotropically. Individual crystals growing in different directions tend to

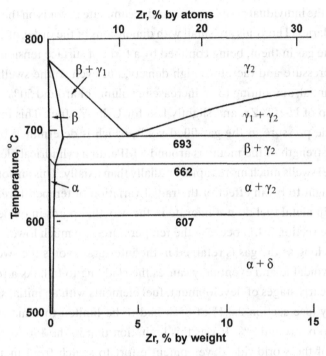

Figure 2.18 The U-Zr phase diagram.

push each other apart by opening large irregular empty voids on the boundaries between them. This process is called "tearing" or "cavitation", and causes the fuel to swell. Uranium fuel at temperatures in the range 400 to 500 °C swells particularly severely, but the presence of plutonium reduces the effect.

Cavitation swelling can be avoided by alloying with a metal such as molybdenum or zirconium that stabilises the γ-phase crystal structure. Figure 2.18 is a part of the uranium-zirconium phase diagram showing the effect of the alloying metal in increasing the temperature range in which the γ-phase is stable. Typical fuel alloys are ternary alloys of U and Pu with 10% Mo or Zr.

Unfortunately the elimination of cavitation only allows another swelling mechanism to operate. Fission-product gas tends to diffuse out of the individual metal crystals and accumulate in voids on the grain boundaries. The voids are small with dimensions of the order of 10 μm and the gas in them, being confined by a form of surface tension, is at high pressure and therefore high density, but even so the swelling is substantial, amounting to an increase in volume of around 50% at 2% burnup of U-10%Zr, and slightly less for U-Pu-Zr fuel. This implies that the pressure in the gas-filled voids, which is determined by the shear strength of the metal, is around 5 MPa. In a cylindrical element the fuel swells much more rapidly radially than axially. This anisotropy is thought to be an effect of the radial variation of temperature. The voids in metal fuel do not migrate in the way that pores in oxide fuel do (see section 2.4.1) because the temperatures are much lower.

As long as the gas is retained in the inter-grain voids the swelling is unavoidable and eventually causes the cladding to fail. As a result, in the early stages of development, fuel elements with an initial smear density (see section 2.3.1) of 85% had to be limited to a maximum burnup of around 3%. It was this limitation that in the 1960s caused most of the worldwide development effort to switch from metal to oxide fuel.

The development in the United States that made it possible to overcome this barrier was to reduce the smear density substantially, to around 75%. This allows a volume increase in excess of 30% before the fuel makes contact with the cladding. At this point the voids on the grain boundaries have grown large enough to join together and release the gas from the fuel matrix into a plenum. Figure 2.19 shows the form of these voids and Figure 2.20 shows the dependence of gas release on the smear density.

Once gas has escaped from the fuel matrix swelling is dramatically reduced but not stopped. The solid fission-product atoms together with some 20% of the gaseous atoms are retained within the metal crystals, resulting in inexorable solid-phase swelling at a rate about half the burnup rate (0.5% increase in volume for each 1% increase in burnup).

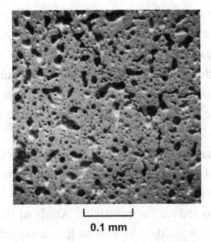

0.1 mm

Figure 2.19 Fission gas voids in irradiated U-Zr fuel.

2.5.3 Mechanical Behaviour during Irradiation

A typical metal fuel element consists of a steel tube 6.0 mm OD, 5.4 mm ID, containing cast U-Pu-Zr pellets initially 4.7 mm in diameter, to give a smear density of 76%. The initial gap between fuel and cladding

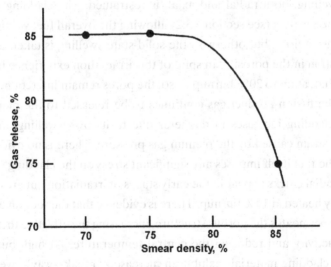

Figure 2.20 The dependence of fission-product gas release on smear density for U-Pu-Zr alloy fuel.

is filled with the sodium in which the whole stack of fuel pellets is immersed. Above and below the core fuel pellets there may be U-Zr breeder pellets, and above the upper breeder there is a gas plenum initially filled with helium and with a volume approximately equal to that of the core fuel. (The option of placing the plenum below the core, which is available for oxide fuel, is precluded by the presence of the sodium.)

If the element is irradiated at a peak linear power of $\sim 50 \, \text{kWm}^{-1}$ to start with the fuel swells rapidly and makes contact with the cladding when the burnup reaches about 2%. The smaller swelling in the axial direction tends to reduce the reactivity. At about the same time as the fuel contacts the cladding its volume has increased by about 30% and its porosity has reached the point at which the majority of the fission-product gas is released to the plenum.

Solid fission products and the small fraction of the fission-product gas that is retained within the crystals continue to accumulate so that the metal continues to expand. The porous fuel is however now weak compared with the cladding with which it is in contact, so further overall swelling, both radial and axial, is restrained. The cladding itself may be swelling (see section 3.3.2) allowing the overall fuel volume to increase a little, but otherwise the solid-state swelling is taken up by reduction in the porosity. In spite of this irradiation experience indicates that, at up to 20% burnup or so, the pores remain interconnected and the fission-product gas continues to be released to the plenum. The cladding increases in diameter due to its own swelling and to creep strain caused by the plenum gas pressure. There is no evidence that the fuel itself imposes any significant stress on the cladding.

Radial cracks appear in the early stages of irradiation but are completely healed at 10% burnup. There is evidence that the sodium eventually permeates the porous structure, increasing the effective thermal conductivity and reducing the central temperature. At high burnup some cladding materials exhibit an increased "break-away" swelling rate and start to swell faster than the fuel so that the gap between fuel

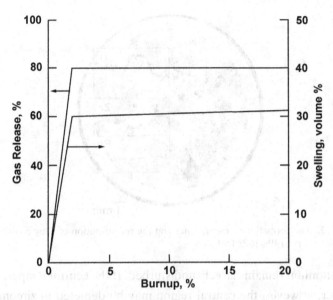

Figure 2.21 The dependence of fission-product gas release and volumetric swelling on burnup for U-Pu-Zr alloy fuel.

and cladding increases. Figure 2.21 shows recommended values of the swelling of the fuel and the fraction of the fission-product gas that is released as functions of burnup for U-Pu-Zr fuel alloy with a wide range of compositions, based on experimental results.

2.5.4 Redistribution of Alloy Components

On irradiation a series of annular zones are formed depending on the radial temperature distribution, as shown in Figure 2.22. There is an outer α-phase zone in the cool periphery with a β-phase zone inside it and, if the temperature is high enough, a γ-phase zone inside that. The redistribution of the constituents of the fuel alloy is also shown in Figure 2.22. The solubility of zirconium is different in the different phases and as a result it tends to migrate out of the β-phase both outwards to the α-phase and inwards to the γ-phase. Uranium is displaced in the opposite directions into the middle annular zone while

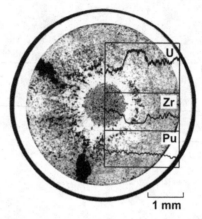

Figure 2.22 The formation of radial zones and the redistribution of alloy constituents in irradiated 71U-19Pu-10Zr fuel.

the plutonium remains largely undisturbed. If the central temperature is lower, however, the central region may be depleted in zirconium. Table 2.3 gives typical local concentrations of the alloy components, based on electron microprobe data, for an element with a high central temperature.

2.5.5 Corrosion of the Cladding

Inter-diffusion between the fuel alloy and the cladding material, sometimes known as "fuel-clad chemical interaction" (FCCI), occurs at the hotter end of the elements. A hard, brittle layer containing rare earth

Table 2.3 *Redistribution of alloy components on irradiation*

		Concentration (w/o)		
		After irradiation to 18% burnup		
Zone	Initial	Inner	Intermediate	Outer
U	71	70	85	58
Pu	19	14	11	18
Zr	10	16	4	24

fission products such as lanthanum, cerium, praseodymium and neo-dymium as well as plutonium, and depleted in nickel is formed on the inner surface of austenitic cladding. It is formed slowly and thicknesses of the order of 100 μm are reported. Ferritic cladding materials form a similar decarburised layer. Because they are weak these layers have to be counted as wastage of the cladding, reducing its effective thickness and thus its ability to support the stress caused by the pressure of the fission-product gas in the plenum.

2.5.6 Reprocessing and Fabrication

The development of metal fuel has taken place in conjunction with that of the "integral fast reactor" (IFR) concept, which is based on close integration of a reactor with a reprocessing and fuel fabrication facility. It is not appropriate to give a full description of the IFR here, but it is necessary to describe it in outline if the nature of the fuel it uses is to be understood.

Central to the IFR system is the reprocessing of the irradiated fuel at high temperature in the molten state, a process called "pyropro-cessing". The earliest version involved "melt-refining". After a brief cooling period the irradiated fuel was melted in a zirconia crucible. Some of the fission products were driven off by evaporation and others formed a residue in the crucible. The fuel, containing the remainder of the fission products and with the addition of new fissile or fertile material as required, was then cast into new fuel elements and sent back to the reactor. The main disadvantage of this simple process was the buildup of fission products after multiple recycling, which reduced the reactivity significantly making it hard to reach high burnup. In addition there were unacceptable losses of uranium and plutonium in the crucible residues.

To eliminate these disadvantages the separation efficiency was increased by replacing the melt-refining process by electro-refining, which involves electrolysis after dissolution in a mixture of molten

Table 2.4 *Energy of formation of chlorides at 500 °C*

Compound	$-\Delta G$ (kJ/g)	Compound	$-\Delta G$ (kJ/g)	Compound	$-\Delta G$ (kJ/g)
$BaCl_2$	367	$CmCl_3$	268	$ZrCl_3$	195
CsCl	367	$PuCl_3$	261	$CdCl_2$	148
RbCl	364	$NpCl_3$	243	$FeCl_2$	122
KCl	362	UCl_3	231	$NbCl_5$	112
$SrCl_2$	354			$MoCl_4$	70
LiCl	345			$TcCl_4$	46
NaCl	339			$RhCl_3$	42
$CaCl_2$	337			$PdCl_2$	38
$LaCl_3$	293			$RuCl_4$	25
$PrCl_3$	288				
$CeCl_3$	287				
$NdCl_3$	284				
YCl_3	272				

salts. The fuel elements are chopped into short lengths and placed in a steel basket that is immersed in a steel vessel containing a molten mixture of lithium and potassium chlorides at 500 °C. At the bottom of the vessel is a layer of molten cadmium. Cadmium chloride is added, and this oxidises the actinides so that they produce a sufficient ion concentration to allow the salt mixture to conduct electricity.

Table 2.4 shows the free energies of formation of the chlorides of the various metals. They fall into three groups: the chlorides of the alkaline earths and most of the rare earths (Ba to Y) are stable and tend to remain in the salt phase; those of the transition metals (Cd to Ru) are unstable so the metals are precipitated in the molten cadmium; whereas the actinides and zirconium form chlorides of intermediate stability that can be separated by electrolysis.

A potential of about 1 volt is applied to the basket of fuel element fragments, which becomes the anode. Two cathodes are used in sequence. The first is a steel rod, on which uranium is deposited as the metal. Plutonium, being more stable, cannot be precipitated until its concentration in the molten salt is high, but when the uranium concentration has been reduced sufficiently it can be precipitated at a second cathode which consists of a crucible containing liquid cadmium. At this cathode, plutonium and the higher actinides form inter-metallic compounds such as $PuCd_6$. After electrolysis the deposits from the cathodes are taken to a furnace where the remaining salts and the cadmium are removed by evaporation. They are then blended to obtain the required fuel composition, melted and cast into moulds to form new fuel pellets.

The decontamination factors are low, by design, and as a result the new fuel, containing significant amounts of fission products, particularly the rare earths, is highly radioactive. For this reason the entire process has to be conducted remotely, but it confers the advantage of protecting the new fuel, and the plutonium it contains, from illicit diversion.

The main advantages of pyroprocessing with electro-refining are that it is cheap and the out-of-reactor fuel inventory is minimised. Another advantage is that the minor actinides are recycled and do not appear in the waste stream (see section 2.7.4). There is evidence that small additions of americium and neptunium to the U-Pu-Zr fuel alloy do not affect its performance adversely, although the high volatility of americium makes for difficulty in the high-temperature fabrication process.

2.6 OTHER FUEL MATERIALS

2.6.1 Carbide

Extensive work has been done on mixed carbide fuel in India in connection with the thorium-based breeder cycle. To initiate this cycle reactors fuelled with ^{239}Pu but with ^{232}Th rather than ^{238}U as the fertile material are needed. In such reactors uranium is essentially redundant, so mixed oxide, which as indicated in section 2.3.3 is limited to about 40% plutonium, or ternary metal alloy with a large uranium

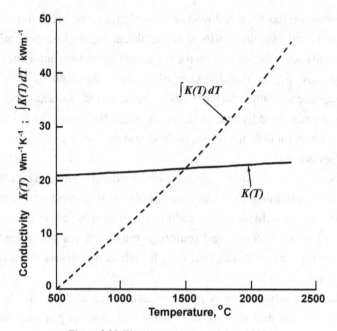

Figure 2.23 The thermal conductivity of UC.

content, are not attractive options. In several respects mixed carbide has superior properties, particularly its high density and high thermal conductivity.

The properties of mixed carbide depend strongly on the stoichiometry. Mixtures of UC and PuC at any ratio form solid solutions. The melting points of UC and PuC are around 2780 °C and 1875 °C respectively, but those of the sesquicarbides U_2C_3 and Pu_2C_3, which are present to a varying degree in a hyperstoichiometric mixture, are 2100 °C and 2285 °C respectively. As a result the melting point varies over a wide range depending on the contents of the mixture.

Figure 2.23 shows the conductivity of UC and its integral. Both are much higher than the corresponding properties of oxide fuel (see Figure 2.1). However the conductivity of PuC is much lower, rising from 10 $Wm^{-1}K^{-1}$ at 700 °C to 20 $Wm^{-1}K^{-1}$ at 1100 °C, and the influence of stoichiometry and impurities such as oxygen on the conductivity is complex. As a result the effective conductivity of a particular fuel material is much lower than that of pure UC, and values of about 7 $Wm^{-1} K^{-1}$ at 500 °C rising to 11 $Wm^{-1} K^{-1}$ at 1000 °C have been reported for $(U_{1-a}Pu_a)C$ fuel with a in the range 0.55–0.7. Nevertheless the conductivity is still much higher than that of oxide, and with a surface temperature of 1000 °C (assuming the gap between fuel and cladding is filled with gas) and a linear heat rating q of 50 kWm^{-1} carbide fuel would have a central temperature of around 1380 °C, well below the melting point. Thus higher values of q without central melting are possible in principle. (It may, however, be that this potential cannot be exploited because of limitations to the rate at which heat can be transported out of the reactor core – see section 3.2.3.)

A disadvantage of carbide fuel is that it carries the risk of carburisation of the steel cladding. The risk depends on the stoichiometry because the carbon activity of $(U,Pu)_2C_3$ is high. Carbide is chemically compatible with sodium so it is possible to fill the gap between fuel and cladding with sodium to provide a good thermal bond. The sodium however tends to transport carbon to the cladding, so helium bonding is usually preferred.

Carbide fuel has a higher density than oxide (because each heavy atom is accompanied by only one light atom rather than two), so it swells more on irradiation. As with metal fuel the swelling is accommodated mainly by a low smear density. At the start of irradiation the fuel pellets form extensive cracks. Swelling closes the gap between fuel and cladding and eventually eliminates the as-manufactured porosity in the pellets. The stress on the cladding from further swelling is relieved, to some extent, by creep of the fuel material. This is particularly important for high-plutonium mixtures, which are softer because the melting temperature is lower.

Because the temperatures are so much lower carbide suffers much less restructuring than oxide (see section 2.4.1) and a smaller fraction of the fission-product gas is released.

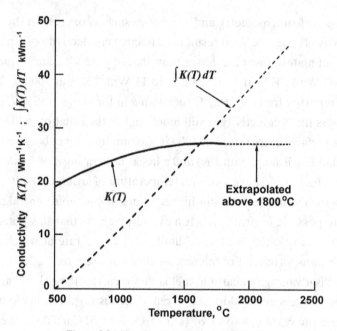

Figure 2.24 The thermal conductivity of UN.

Carbide is usually produced by carbothermic reduction of oxide. Oxide and carbon powders are mixed, formed into pellets and heated, and the oxygen is driven off in the form of carbon monoxide. Further heating in a hydrogen atmosphere can control the stoichiometry and remove residual oxide and other impurities. A significant disadvantage of carbide is that it is pyrophoric, so the manufacturing process and all fuel-handling activities have to be carried out in an inert atmosphere.

2.6.2 Nitride

The behaviour of nitride fuel is very similar to that of carbide. The thermal conductivity of UN is shown in Figure 2.24, showing that, as in carbide, high linear heat ratings are possible in principle. The conductivity of PuN is about half that of UN and for mixtures containing 20 to 35% Pu values in the range of 20 $Wm^{-1} K^{-1}$ are possible (depending on the porosity). There are fewer complications due to variation in stoichiometry because higher nitrides of uranium are unstable and can be dissociated to UN by heating to 1400 °C, and plutonium forms only PuN. The melting point of UN is about 2740 °C and PuN dissociates at about 2570 °C.

Nitride can be produced, like carbide, by carbothermic reduction of oxide, but in this case in a nitrogen atmosphere. The disadvantage of this method is that the product contains carbon and oxygen as impurities, and these pose some risk of carburisation of the cladding. Nitride is not pyrophoric.

Under irradiation in general nitride behaves similarly to carbide, but it swells less and releases less fission-product gas. It is also more prone to cracking.

The principal disadvantage of nitride fuel made with atmospheric nitrogen is the reaction $^{14}N(n,p)^{14}C$. The ^{14}C produced during irradiation is a hazard during reprocessing and is an unwanted additional component of the radioactive waste that has to be disposed of. It could in principle be reduced or eliminated by enriching the nitrogen from which the fuel is made in ^{15}N (which constitutes 0.36% of natural nitrogen) but this would be expensive, and it would be difficult to conserve it during reprocessing.

2.7 FUEL FOR CONSUMER REACTORS

2.7.1 Consumption of Plutonium

The preceding part of the chapter has been written with breeder reactors in mind. Different fuel is needed for a reactor designed to consume plutonium rather than breed it, the principal difference being that it should not contain uranium or at least that it should have a minimum of it. There is little irradiation experience with high-Pu fuels so in what follows it is possible only to indicate what seems most likely to be of use in practice.

If plutonium is to be consumed then it is clearly desirable that it should be consumed as quickly as possible. Plutonium is consumed by fissioning it, which produces energy. The power of a plutonium-consuming reactor should therefore be high if it is to be effective. A 2500 MW (heat) reactor with an 80% load factor, fuelled with pure plutonium with no breeding, would consume about 800 kg of plutonium per year.

The constraints of heat transport (see section 3.2) mean that a 2500 MW core would have to have a volume of about 3 m^3 and would need some 1700 kg of plutonium to make it critical. If the plutonium were in fuel elements generating an average of 32 kWm^{-1} (i.e. a maximum rating of 50 kWm^{-1} at the centre of the core) the fuel elements would have to have a total length of 78 km. If the fuel were PuO$_2$ with an 80% smear density the radius of the pellets would be about 1 mm, and if it were metal or some other ceramic the radius would be even smaller! Clearly this is impossibly small for practical fuel elements: they have to be larger, and in practice a diameter of about 6 mm is the minimum that is economically and structurally possible. Thus the plutonium has to be diluted with some inert material, which in effect is present to replace the uranium that makes up such a large part of the fuel of a breeder reactor.

There are two ways in which a diluent material can be incorporated in a ceramic fuel. It can either be a solid solution in the ceramic, or a second component in a two-component mixture. Because of the difficulty of dissolving PuO$_2$ in nitric acid for purex reprocessing, and also because it reacts chemically with sodium, plutonium oxide solid solutions such as (Pu,Zr)O$_2$ are not attractive. The most promising ceramic solid solution is (Pu,Zr)N. There are few data on its physical properties but theoretical estimates suggest that, as in the case of Pu-Zr alloy, it would perform well in a 50 kWm^{-1} fuel element.

In the 1960s, when a solution to the problem of fuel swelling was being sought, "cermet" fuels were investigated. A cermet is a sintered mixture of ceramic and metal powders. The idea was that the metal would form a strong matrix that would constrain the swelling of the fuel ceramic. Some irradiation testing was done and high burnups of UO_2-stainless steel cermets were achieved. However, the project was abandoned when it was realised that if the metal fraction was high enough to prevent swelling it would absorb so many neutrons that the breeding ratio would be reduced unacceptably. This of course would not be a disadvantage in a plutonium-consuming reactor, and cermets of PuO_2 with steel or other metals such as chromium, vanadium or tungsten may be attractive.

By analogy with a cermet, a sintered mixture of two mutually insoluble ceramic powders is sometimes called a "cercer". The range of suitable diluent ceramics is restricted because the resulting cercer would have to be soluble in nitric acid if the fuel is to be reprocessed by the purex process. Cercers of PuO_2 and magnesia (MgO) or yttria (Y_2O_3) are possibilities. Ceria (CeO_2) should probably be ruled out because it is incompatible with liquid sodium and would swell severely in the event of a cladding failure.

Because of the success of U-Pu-Zr alloy fuel Pu-Zr alloy appears attractive. The melting point of plutonium is very low at 640 °C but that of zirconium is 1852 °C. The *solidus* temperatures of Pu-Zr containing 20% and 40% Pu are about 1680 °C and 1480 °C respectively. Since their conductivities (without porosity) are around $20 \ Wm^{-1} \ K^{-1}$, comparison with the 70U-20Pu-10Zr fuel of section 2.5.1 seems to indicate that a high-Pu metal-fuelled element would have acceptable irradiation performance at a linear rating of $50 \ kWm^{-1}$.

2.7.2 Consumption of Higher Actinides – Ceramic Fuel

As explained in section 1.4.1 isotopes of americium and curium are produced in all uranium-fuelled reactors, in greater quantities the higher the burnup of the fuel. They constitute hazardous waste products and there may be an incentive to eliminate them. Because they are fissionable this can be done in principle by incorporating them as fuel in a fast reactor. However if the objective is to reduce their quantity the reactor should not generate more of them, and therefore, as in the case of a plutonium-consumer reactor, the fuel should contain little or no uranium and would have to be diluted.

Higher-actinide ceramic fuel would probably be reprocessed in dedicated facilities, separate from the mainstream, using either the purex process or a modification of it (see section 2.3.3) with lower separation efficiency and a product stream matched to the fuel-cycle requirements. For this reason the fuel and any diluent materials must be soluble in nitric acid.

In addition the higher actinide fuel would have to be compatible with the coolant in the sense that, in the event of a small cladding failure, gross swelling or any other reaction that would cause the release of significant quantities of fuel material or fission products to the coolant would have to be impossible. This limits the choice of fuel material for a sodium-cooled reactor, but not necessarily for reactors cooled with gas or lead. In principle the fuel material could be a pure compound, a solid solution of fuel and a diluent compound, or a two-phase mixture of fuel and a diluent compound. The two-phase mixture could be either a cermet or a cercer.

Information for most of the candidate materials is incomplete. Physical property data on thermal conductivity or melting point are sometimes known or can be inferred, but there is usually nothing on properties such as thermal creep. There is little irradiation experience but some aspects of the behaviour under irradiation can be deduced from the phase diagram. (If there are phase changes in the operating temperature range the structure is unlikely to be stable.) There is often no empirical information on the compatibility with cladding or coolant, although theoretical inferences can sometimes be made. Information on curium compounds is scarce so the selections are mostly based on americium data.

The most obvious choices for pure compound fuel materials are summarised in Table 2.5. The most important considerations are that

Table 2.5 *Minor actinide fuel materials*

Material	Disadvantages	Comments
AmO_2	Reacts with Na. High O potential. Low thermal conductivity.	Vaporises above 1400 °K.
Am_2O_3	Phase changes.	Inert matrix may stabilise structure.
AmN		Volatile. Produces ^{14}C.
Am_2C_3	Pyrophoric	May cause embrittlement of cladding.

carbides are pyrophoric, AmO_2 has a high oxygen potential, and Am_2O_3 has several phase changes.

Table 2.6 summarises the main candidate diluent materials for cermets or cercers. Nitrides might require enrichment in ^{15}N to avoid production of ^{14}C. In addition to the materials listed in the table Si_3N_4, TiN, YN, and AlN are also candidates.

2.7.3 Preferred Ceramic Fuel Materials for a Consumer of Higher Actinides

The most stringent requirements are those for selecting fuel for a dedicated consumer reactor fuelled with minor actinides alone, mainly because the linear rating of the fuel has to be as high as in a conventional reactor. Both the thermal conductivity and the melting point are lower for minor actinide oxides than for UO_2 or PuO_2. This rules out single-component oxide fuel. Similarly the thermal conductivity of americium-bearing oxide cercers is so low that realistic linear ratings cannot be achieved. The situation is better for oxide cermets as far as linear rating is concerned, but unless the metal fraction in the cermet is to be kept around 50 v/o or more it cannot be expected to seal the fuel, so in the event of cladding failure there is a risk of swelling on contact with Na. Thus oxide in any form is effectively ruled out for a dedicated

Table 2.6 *Minor actinide diluent materials*

Material	Disadvantages	Comments
MgO		
Al_2O_3	Swells on irradiation	Insoluble in nitric acid.
Y_2O_3		
CeO_2	Incompatible with Na	
$MgAl_2O_4$		Insoluble in nitric acid.
$Y_3Al_5O_{12}$		Insoluble in nitric acid.
$CePO_4$	Swells on irradiation. Incompatible with Na.	May be incompatible with steel cladding.
$ZrSiO_4$	Swells on irradiation. Incompatible with Na.	May be incompatible with steel cladding.
SiC	Incompatible with steel cladding.	
ZrN		Forms solid solution with Am or Pu.
CeN	May swell on irradiation	
Cr metal		Forms eutectic with AmN. May not be soluble.
V metal		Forms eutectic with AmN.
Steel		Forms eutectic with AmN at ~1400 °C.

sodium-cooled consumer, although oxide cermets might be possible in consumers cooled by lead or gas. The remaining possibilities are

- Pure compound   (Am,Cm)N
- Solid solutions   (Am,Cm,Zr)N or (Am,Cm,Y)N
- Cercers   (Am,Cm)N+TiN or (Am,Cm)N+AlN
- Cermets   (Am,Cm)N+W

An alternative to a reactor fuelled entirely with minor actinides is a "normal" plutonium-fuelled reactor with special fuel subassemblies containing the minor actinides in or around the core. The power rating of these "target" assemblies would not have to be so high, especially if they were placed round the periphery of the core, so they could make use of oxide fuel.

Pure minor actinide oxides are ruled out because they are unlikely to be stable under irradiation, but it may be possible to stabilise them in solid solution with Zr or Y to give a fuel suitable for targets. Similarly the low thermal conductivity of Am-bearing oxide cercers may be acceptable. The lower rating of the targets might also make cladding failure sufficiently unlikely that swelling of the fuel material on contact with sodium is less important, allowing oxide cermets or cercers to be considered. The possibilities for target fuel include all those identified above and in addition the following:

- Solid solutions $(Am,Cm,Zr)O_2$ or $(Am,Cm,Y)_2O_3$
- Cercers $(Am,Cm)O_2+MgO$
- Cermets $(Am,Cm)O_2+W$, Cr, V or steel

2.7.4 Consumption of Higher Actinides – Metal Fuel

As explained in section 2.5.6 one of the advantages of the IFR system is that the higher actinides do not appear as waste products but are "automatically" recycled and consumed along with the plutonium. Irradiation testing has shown that the performance of metal fuel is not significantly affected by the presence of higher actinides. Figure 2.25 shows the radial redistribution of the alloy constituents. Neptunium does not migrate whereas americium tends to follow the zirconium and to be precipitated in pores.

Higher actinides produced in thermal reactors could be consumed in this way if, in the reprocessing plant, they were not separated from the plutonium but used with it as a feed for a metal-fuelled IFR reactor. However if existing higher actinide waste inventories were to be consumed fuel with much higher concentrations of americium and curium would be needed.

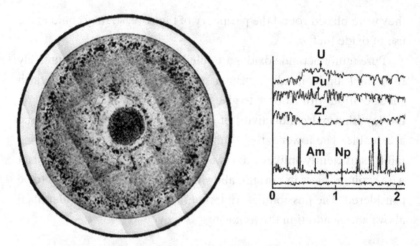

Figure 2.25 The formation of radial zones and the redistribution of alloy constituents in irradiated 67.5U-20Pu-1.3Np-1.2Am-10Zr fuel.

REFERENCES FOR CHAPTER 2

Carmac, W. J., D. L. Porter, Y. I Chang, S. L. Hayes, M. K. Mayer, D. E. Burkes, C. B. Lee, T. Mizuno, F. Delage and J. Somers (2009) Metallic Fuels for Advanced Reactors, *Journal of Nuclear Materials*, 392, 139–150

Chamberlain, A., H. A Taylor, R. H. Allardice and J. A. Gatley (1978) The Optimisation of Fuel Design in relation to Reactor Performance and the Conflicting Demands of other parts of the Fuel Cycle, *Optimisation of Sodium-Cooled Fast Reactors*, pp 133–136, British Nuclear Energy Society, London

Chang, Y. I. and C. E. Till (2011) *Plentiful Energy: The Story of the Integral Fast Reactor*, CreateSpace online publishing

Cox, C. M. and F. J. Homan (1970) Performance Analysis of a Mixed-Oxide LMFBR Fuel Pin, *Nuclear Applications and Technology*, 9, 317–325

Crawford, D. C., D. L. Porter and S. L. Hayes (2007) Fuels for Sodium-Cooled Fast Reactors: US Perspective, *Journal of Nuclear Materials*, 371, 202–213

Findlay, J. R. (1974) The Composition and Chemical State of Irradiated Oxide Reactor Fuel Material, pp 31–39 in *Behaviour and Chemical State of Irradiated Ceramic Fuels*, IAEA, Vienna

Hofman, G. L., L. C. Waters and T. H. Bauer (1996) Metallic Fast Reactor Fuels, *Progress in Nuclear Energy*, 31, 83–110

International Atomic Energy Agency (2009) *Status of Minor Actinide Fuel Development* Technical Report NF-T-4.6, IAEA, Vienna

International Atomic Energy Agency (2011) *Status and Trends of Nuclear Fuels Technology for Sodium-Cooled Fast Reactors* Technical Report NF-T-4.1, IAEA, Vienna

Kim, Y. S. and G. L. Hofman (2003) *AAA Fuels Handbook*, Argonne National Laboratory, Argonne, Illinois, USA

Kittel, J. H., B. R. T. Frost, J. P. Mustelier, K. Q. Bagley, G. C. Crittenden and J. Van Deivoet (1993) History of Fast Reactor Fuel Development, *Journal of Nuclear Materials*, 204, 1–13

Meyer, R. O., D. R. O'Boyle and E. M. Butler (1973) Effect of Oxygen-to-Metal Ratio on Plutonium Redistribution in Irradiated Mixed-Oxide Fuels, *Journal of Nuclear Materials*, 47, 265–267

Olander, D. R. (1976) *Fundamental Aspects of Nuclear Reactor Fuel Elements*, Energy Research and Development Administration, Washington, DC

Pahl, R. G. and R. S. Wisner (1990) Steady-State Irradiation Testing of U-Pu-Zr Fuel to >18 at% Burnup, *Proceedings of the International Conference on Fast Reactor Safety IV*, American Nuclear Society, Hinsdale, Illinois, USA

Perrin, J. S. (1972) Effect of Irradiation on Creep of UO_2-PuO_2, *Journal of Nuclear Materials*, 42, 101–104

Powell, H. J. (1974) Fission Product Distribution in Fast Reactor Oxide Fuels, pp 379–392 in *Behaviour and Chemical State of Irradiated Ceramic Fuels*, IAEA, Vienna

Rodriguez, P. (1999) Mixed Plutonium-Uranium Carbide Fuel in Fast Breeder Test Reactor, *Bulletin of Materials Science*, 22, 215–220

Zegler, S. T. (1962) *The Uranium-Rich End of the Uranium-Zirconium System* Report ANL-6055, Argonne Nuclear Laboratory, Argonne, Illinois, USA

3

REACTOR CORE

3.1 INTRODUCTION

Having described the neutron physics of a fast reactor in Chapter 1 and the behaviour of the fuel elements in Chapter 2, in this chapter we discuss the engineering of the core of a power-producing fast reactor. The three following sections deal with heat transfer, materials and structure.

Heat transfer comes first because the dimensions of the fuel elements and of the core are determined mainly by the demands of heat transfer. The fuel elements have to be of the right dimensions to transfer heat to the coolant at the required rate without overheating. The core has then to be large enough to allow enough coolant to flow through it to take the heat away. This section deals mainly with liquid metal coolants.

Once the main dimensions are fixed the form of the core is determined by the properties of the materials of which it is made, and in particular by the way these properties are affected by neutron irradiation and exposure to the coolant. The structure of the core has then to be designed within these constraints to hold the fuel in place, to allow it to be changed when necessary, to distribute the coolant flow correctly, and to provide for the control rods.

150

3.2 HEAT TRANSFER AND TRANSPORT

3.2.1 Fuel Element Rating

The primary economic pressure is to maximise the power output from a reactor, because this gives the best return on the capital invested in the reactor plant and the inventory of fuel committed to the reactor and the reprocessing cycle, and also maximises the breeding of fissile material or the consumption of waste products, whichever is required. Whatever the purpose of the reactor the heat generated has to be removed from the reactor core, and the power is limited by heat-transfer considerations. The crucial limits are set by conduction of heat within the fuel elements and by the flow of coolant through the core.

As explained in Chapter 2 in power reactors the fuel elements are in the form of long tubes of cladding, usually steel, containing the fuel in the form of ceramic or metal pellets or powder. If the power density in the fuel material due to fission is $Q \, \mathrm{Wm^{-3}}$, then q, the linear rating of the fuel element, is given by $q = \pi R_f^2 Q$, and if ΔT_f is the temperature difference between the centre and the surface of a cylindrical fuel pellet, in the case of constant thermal conductivity, q is given by

$$q = 4\pi K_f \Delta T_f. \tag{3.1}$$

As we have seen in Chapter 2 for most fuel materials the maximum acceptable value of q is about $50 \, \mathrm{kWm^{-1}}$. If q is fixed the power density $Q = q/\pi R_f^2$ can in principle be increased indefinitely by reducing the radius of the fuel, but a practical limit is set by the cost of manufacture which rises rapidly for very small fuel elements. For this reason the fuel radius cannot be less than about 2.5 mm, which limits the maximum power density in the fuel to about $2.55 \, \mathrm{GWm^{-3}}$.

3.2.2 Distribution of Power Density

While the reactor is operating the heat transferred from the fuel arises almost entirely from fission. Q is given by

$$Q \approx \Delta E_f \sum_g \phi_g \Sigma_{fg}^{fuel},$$ (3.2)

where ΔE_f is the difference between the internal energy of the reactants and products of a fission event, assuming there is no temperature change and that the products are at rest. ΔE_f is about 200 MeV, or 3.2×10^{-11} J, for ^{239}Pu and is only slightly different for other isotopes. Σ_{fg}^{fuel} is the group fission cross-section for the fuel material. It is not the same as Σ_{fg} in equation 1.9, which is an average cross-section for a region of the reactor including structure and coolant.

Equation 3.2 is not exact. Some of the energy is transferred by neutrons and radiation and appears in the structure and coolant and even in the shielding. Some is transferred when radioactive fission products decay and so appears after the fission has taken place. While the reactor is operating γ-heating in the structure, the outer parts of the breeder and the shield is important, and when it is shut down radioactive decay heating in the fuel is important. Nevertheless during operation some 97% of the energy appears promptly in the fuel.

The neutron spectrum is nearly the same throughout the core so Q is roughly proportional to the total flux, the distribution of which is shown in Figure 1.13. The maximum value, Q_{max}, occurs at the centre of the core or at the inside of the outer enrichment zones. The average Q along the most highly rated fuel element is lower than Q_{max} and the average Q for the whole core is lower still.

To take account of the variation of power density across the core "form factors" are defined. An axial form factor f_z is given by

$$f_z = \int_0^H Q(z)dz / H Q_{max},$$ (3.3)

where $Q(z)$ is the value of Q at a distance z from the bottom of the most highly rated fuel element and H is the height of the core. The radial form factor, f_r, is then defined by

$$f_r f_z = \int_{core} Q \, dv / V Q_{max}, \qquad (3.4)$$

where the integral runs over the whole core and V is its volume. Both f_r and f_z are usually about 0.8 and the average power density over the whole core is about 0.64 of the maximum. The power density at the ends of the most highly rated element is about 0.4 Q_{max}, and at the extremities of the core it is as low as 0.2 Q_{max}.

If the peak linear rating is 50 kWm^{-1} therefore the average for the most highly rated element is about 40 kWm^{-1} and the average over the whole core is about 32 kWm^{-1}, whereas the least highly rated element has an average linear rating of about 18 kWm^{-1}. If a reactor is to produce 2500 MW of heat it requires a minimum of about 78 km of fuel elements in total, whatever their radius. If the fuel radius is 2.5 mm the total volume of fuel is at least 1.53 m^3.

3.2.3 Heat Transport from the Core

Two other important dimensions, the height of the core and the spacing between the fuel elements, are determined mainly, although not completely, by the coolant. The flow of coolant through the core is subject to limitations on temperature rise, pressure drop, and velocity, none of which can be too high.

If the height of the core is H, so that the power output from the highest rated fuel element is $q_{max} f_z H$, then

$$q_{max} f_z H = A v_{max} \rho c \Delta T_c, \qquad (3.5)$$

where ρ and c are the density and specific heat capacity of the coolant, ΔT_c is the temperature rise of the coolant as it passes through the core, and v_{max} is the mean velocity of the coolant associated with the highest rated element. A is the coolant flow area per fuel element and depends on the spacing of the fuel elements.

The temperature rise ΔT_c is the difference between the coolant temperatures at the core outlet and inlet. The outlet temperature is fixed by the need for adequate strength and resistance to creep in the cladding and structural materials, whereas the inlet temperature is determined by the design of the steam plant and by the need for adequate resistance to thermal shock in the event of a sudden change such as a turbine trip (see section 4.2.4).

The maximum coolant velocity v_{max} is limited by considerations of erosion, vibration and pressure drop. For sodium it cannot exceed about 10 ms^{-1}, partly because of the risk of erosion of steel cladding, and partly because vibration of fuel elements and structural components is much more difficult to control at higher velocities. For lead, with its higher density, the limit is much lower and is usually set at 3 ms^{-1} or less.

As explained later (section 3.4.1) it is normal to restrict the coolant flow to less highly rated fuel elements so that the temperature rise is uniform across the core. The pressure drop is therefore determined by the most highly rated fuel. It is convenient to think of it as being given by an expression of the form

$$\Delta P_c = C \left(\tfrac{1}{2}\rho v_{max}^2\right) (4H/D_H), \tag{3.6}$$

where D_H is the hydraulic diameter of the coolant channels (which depends on the separation of the fuel elements) and C is a constant with the nature of a friction factor. Equation 3.6 is useful as an illustration but in reality the situation is not so simple because the value of C depends on the details of the design of the coolant channels and especially on the nature of the fuel element supports.

If the radius of the fuel elements is fixed then A in equation 3.5 and D_H in equation 3.6 are determined by the fuel element spacing. Consequently these two equations can be used to determine v_{max} and ΔP_c as functions of H and the spacing. Typical results for a 1 m high

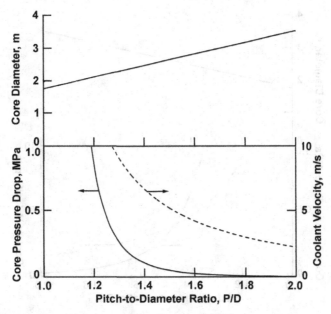

Figure 3.1 The dependence of coolant velocity, pressure drop and core diameter on fuel-element pitch-to-diameter ratio for a sodium-cooled reactor.

core $(H = 1)$ cooled with sodium are shown in Figure 3.1. The fuel elements are taken to be arranged in a triangular array the shape of which is determined by the ratio of pitch to diameter P/D, and the power is taken to be 2500 MW (heat). P/D determines the coolant flow area associated with each fuel element and, because the total number of fuel elements is fixed by the total power output and the linear rating, the overall diameter of the core.

The choices open to the designer can now be seen. If for example ΔP_c is required to be no greater than 0.3 MPa, P/D must be greater than about 1.3 and the core diameter something more than 2 m. If the core is lower $(H > 1 \text{ m})$ the pressure drop and coolant velocity are lower so P/D can be reduced, but at the cost of making the core diameter greater. This is undesirable becauseit forces the entire reactor structure to be larger and increases the capital cost of the plant. Higher

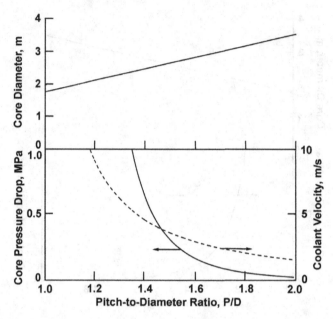

Figure 3.2 The dependence of coolant velocity, pressure drop and core diameter on fuel-element pitch-to-diameter ratio for a lead-cooled reactor.

values of ΔP_c are possible, but if it exceeds about 700 kPa cavitation and the generation of noise may be important.

Figures 3.2 and 3.3 illustrate the choices available with different coolants. If lead is used its high density requires a much higher pressure drop, and this and the more stringent limitation on velocity to avoid damaging erosion force a considerably larger value of P/D, typically at least 1.5, and with it a larger core diameter. Velocities are much greater if the coolant is a gas, of course, and even with the relatively high density of supercritical CO_2 at a pressure of 20 MPa, to which Figure 3.3 refers, they are of the order of 50 ms^{-1}, but the pressure drop is not unduly high and $P/D < 1.4$ is possible.

There are many other choices to be made because other parameters, such as the maximum linear rating and the core temperature rise, can be varied. The ultimate choice depends on the criteria used for optimisation, including safety (see Chapter 5) and economics.

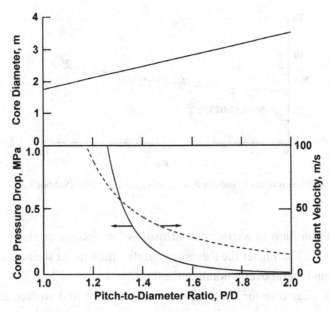

Figure 3.3 The dependence of coolant velocity, pressure drop and core diameter on fuel-element pitch-to-diameter ratio for a CO_2-cooled reactor.

3.2.4 Heat Transfer to the Coolant

Liquid metals are used as coolants because they are very good media for heat transfer. The high thermal conductivity is due to the motion of the electrons in the conduction energy band, which are much more mobile than the positive ions. Being so light the electrons transport energy but little momentum so the thermal conductivity is high but the viscosity is not. As a result the Prandtl number $Pr = \mu c / K$ (where μ is the viscosity, c the specific heat capacity and K the thermal conductivity), which for most fluids is of the order of one because energy and momentum are both transported by the diffusion of molecules, is very small for liquid metals. For sodium at 600°C, $Pr = 4.2 \times 10^{-3}$ and $K = 62.3 \ \mathrm{Wm^{-1}K^{-1}}$, whereas for water at 100°C, $Pr = 1.72$ and $K = 0.68 \ \mathrm{Wm^{-1}K^{-1}}$.

This does not mean, however, that the heat is transferred 100 times more readily to sodium than to water because in a flowing liquid

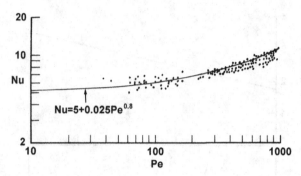

Figure 3.4 Heat transfer to sodium flowing in a cylindrical tube (Subbotin *et al* 1963).

the bulk motion as well as the diffusion of molecules or electrons is important. The higher the Reynolds number the smaller the advantage of a liquid metal over a nonmetallic fluid.

The heat transfer coefficient h between a heated surface and a fluid is defined by $h = Q/\Delta T$, where Q is the local heat flux (Wm^{-2}) and ΔT is the difference between the local temperature of the heated surface and the mixed mean fluid temperature. For fluid flowing in a channel h can be nondimensionalised conveniently in the Nusselt number $Nu = hD_H/K$, where D_H, is the hydraulic diameter of the channel. Dimensional analysis then shows that for a liquid metal, in which turbulence does not make a significant contribution to heat transfer so that h is independent of μ, $Nu = Nu\,(Pe)$, where Pe is the Peclet number $\rho c v D_H/K$, ρ is the fluid density and v is the mean fluid velocity.

Figure 3.4 shows the variation of Nu with Pe for sodium flowing in a cylindrical tube. As $Pe \to 0$, Nu tends to a constant value. If the liquid velocity were uniform across the tube and heat transfer were purely by conduction Nu would be 8. The effect of the variation of velocity across the tube is to reduce Nu, whereas turbulence increases Nu at high Pe. A good fit to the experimental data is

$$Nu = 5 + 0.025Pe^{0.8}. \tag{3.7}$$

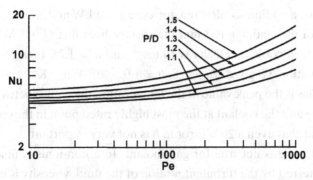

Figure 3.5 Heat transfer to a liquid metal flowing axially in a triangular array of tubes.

For flow at high Reynolds numbers parallel to a triangular array of cylindrical rods various correlations such as

$$Nu = 4.0 + 0.0063\alpha^{3.8}Pe^{0.86} + 0.16\alpha^{5.0}, \tag{3.8}$$

where α is the pitch-to-diameter ratio P/D, have been proposed. Nu increases with α as shown in Figure 3.5, but h does not increase in the same way because D_H also increases with α.

The experimental heat transfer data for liquid metals, especially for arrays of tubes, are in most cases very scattered, and the experimental values of Nu usually lie below, and sometimes a factor of two below, the theoretical predictions. This is due in part to the difficulty of doing the experiments. If for example $Nu = 8$ for sodium flowing in a tube 10 mm in diameter, then $h = 5 \times 10^4$ Wm^{-2}K^{-1}. Even if the heat flux is as high as 100 kWm^{-2} the temperature difference is only 2 K and cannot be measured accurately. In addition the temperature difference can be affected considerably by even very slight contamination of the heater surface. The presence of oxide films may explain why experimental values of Nu for sodium are frequently lower than those for mercury, for example.

What is true for the experiment is also true for the reactor, and in many cases uncertainty in Nu is not important because the temperature difference is so small. If the linear rating q of a fuel element at the centre of a sodium-cooled reactor core is 50 kW m^{-1} and the outer radius of the cladding is 3 mm the surface heat flux Q is 5 MWm^{-2}. If the mean coolant velocity is 10 ms^{-1} and $\alpha = 1.25$, then $Pe = 353$ and equation 3.8 gives $Nu = 6.8$, $h = 1.0 \times 10^5$ Wm^{-2}K^{-1} and $\Delta T = $ 14 K. This is the peak value of the temperature difference between the cladding and the coolant at the most highly rated point in the core and it is clear that even a 20% error in h is not very important.

The same is not true for gas coolant. In a non-ionised fluid heat is transferred by the turbulent motion of the fluid, viscosity is important, and heat transfer depends strongly on the Reynolds number $Re = \rho v D h/\mu$. The Nusselt number is usually given by an empirical relationship, a simple example of which is

$$Nu = 0.023\, Re^{0.8} Pr^{0.4}. \tag{3.9}$$

Taking again the example of CO_2 at 20 MPa, if it were flowing at $60\,\text{ms}^{-1}$ over fuel elements of diameter 6 mm with $P/D = 1.3$, Re would be about 10^6 and Nu about 1.3×10^3, giving a heat transfer coefficient h of 1.7×10^4 Wm^{-2}K^{-1}. This, with $Q = 5$ MWm^{-2}, would give a temperature difference between the cladding and the coolant ΔT of about 300 K, compared with 14 K for sodium under comparable conditions.

3.2.5 Coolant and Cladding Temperatures

The mean coolant temperature $T_c(z)$ at a height z above the bottom of the core, neglecting heat generated in the axial breeder, is

$$T_c(z) = T_{ci} + \int_0^z q(z)dx/Av\rho c, \tag{3.10}$$

where T_{ci} is the coolant inlet temperature. The mean cladding surface temperature $T_s(z)$ is

$$T_s(z) = T_c(z) + q(z)/2\pi R_s h, \tag{3.11}$$

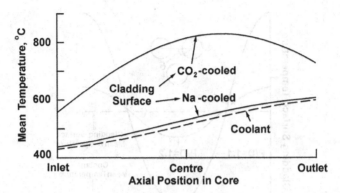

Figure 3.6 Coolant and cladding mean temperatures.

where R_s is the outer radius of the cladding. Typical variations of T_s and T_c with z for sodium and gas coolants are shown in Figure 3.6. For liquid metal coolants T_s follows T_c closely and the maximum value of T_s occurs at the top of the core, but for gas for which h is much smaller T_s reaches its maximum near the centre of the core.

The maximum cladding temperature is a very important limit on the design of the reactor. It is set principally by the creep strength that the cladding has to have, and in turn it sets the coolant outlet temperature and hence influences the thermodynamic efficiency of the generating plant. The maximum cladding temperature is, however, not the same as the maximum value of T_s because for a number of reasons there are, or may be, local variations that give locally high cladding temperatures or "hot spots". There are many possible causes of the irregularities in the heat transfer that give rise to hot spots, of which the most important are as follows.

The variation of coolant temperature across the coolant channel between three adjacent fuel elements is significant and is reflected as a variation of the cladding temperature round a fuel element. T_s is higher at a point near a neighbouring element and lower opposite a gap between elements, as shown in Figure 3.7.

The variation is larger for close-packed arrays and for fluids with low Prandtl numbers. This is because heat transfer within the coolant

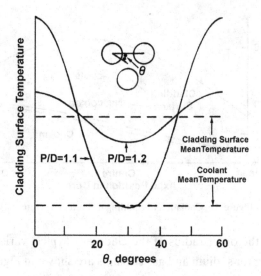

Figure 3.7 The variation of cladding temperature round a fuel element.

(i.e. from the intensely heated gap between two fuel elements to the centre of the subchannel between three fuel elements where most of the coolant flows) is strongly affected by the coolant velocity and is therefore relatively insensitive to Pr. Heat transfer across the boundary layer depends mainly on conduction and therefore varies strongly with Pr. If Pr is high (i.e. close to 1), boundary-layer heat transfer is poor and the cladding surface is in effect partially insulated from the temperature variation in the coolant. The temperature variation around a fuel element in a gas-cooled reactor is thus less severe than in a liquid-metal-cooled reactor with the same linear heat rating.

The coolant channels near the wrapper are often larger than those in the rest of the subassembly so that the edge fuel elements are overcooled. Coolant and cladding temperatures in the edge channels are lower than average, implying that the cladding temperatures in the centre of the subassembly are higher than average, giving another form of hot spot.

Except at the centre of the core the power density varies across a subassembly and may be some 10% higher on one side than the other.

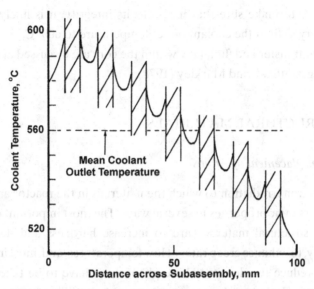

Figure 3.8 The coolant temperature distribution at the outlet from a 169-element sodium-cooled subassembly.

As a result coolant and cladding temperatures on the hot side are higher than the average. The coolant temperature distribution across the outlet of a typical subassembly is shown in Figure 3.8 which illustrates the effects of "power tilt", the overcooling of edge fuel elements, and the variation of temperatures across each coolant channel.

The fuel elements are held in position by grids or wire wraps as explained in section 3.4.1. Either form of support modifies the coolant flow in places and although the effect is usually to increase the turbulence locally and hence increase h and decrease T_s there is a possibility of a region of poor cooling. Other hot spots occur at random because of variations of the dimensions and composition of various components within the manufacturing tolerances.

Cumulatively the allowances that have to be made for all these effects are very significant and may amount to as much as half the mean coolant temperature rise through the core, something in the range of 70–100 K. Thus if the cladding temperature T_s has to be limited to 700 °C to make sure that it retains its integrity, it is likely to be necessary to limit the coolant outlet temperature to 600 °C.

Heat transfer and fluid flow within the core are discussed at length by Tang, Coffield, and Markley (1978).

3.3 STRUCTURAL MATERIALS

3.3.1 Displacement of Atoms

The neutron irradiation to which the materials in the reactor are subject alters their properties in several ways. The most important effects on the structural materials are to increase hardness and decrease ductility, to enhance creep rates at low temperatures, and, most important, to reduce the density. These phenomena have to be taken into account in design of the reactor along with familiar effects such as thermal creep, fatigue and corrosion.

Irradiation affects the properties of non-fissile materials in two ways. Neutron scattering interactions displace atoms from their sites in the crystal lattice, creating vacancies and interstitial atoms in equal numbers, and neutron absorption by (n,α) and (n,p) interactions creates atoms of helium, hydrogen and other transmutation products within the crystals. Helium has the greatest effect on the properties of the material.

A useful way to characterise the extent of the irradiation received by a piece of material is to specify the "displacement dose", which is the average number of times an atom has been displaced from its lattice site. Each elastic scattering interaction imparts kinetic energy E_p to the target nucleus, where E_p is a random variable distributed uniformly (if the scattering is isotropic) in the range $0 - \mu E_n$, where E_n is the neutron energy, $\mu = 4A/(1+A)^2$, and A is the atomic weight of the target. For iron $\mu = 0.069$, so a 1 MeV neutron can impart up to 69 keV to an iron nucleus from which it is scattered elastically.

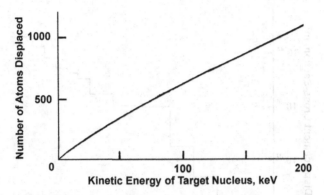

Figure 3.9 The number of iron atoms displaced by a nucleus that is the target of a neutron scattering event (TRN model).

It requires only some 25 eV to displace an iron atom from its site in the crystal lattice, so the target nucleus may well have enough energy to displace several hundred atoms as it moves through the crystal. Some of the kinetic energy is however taken up by inelastic scattering interactions, and the number of atoms displaced by elastic scattering depends on the direction in which the target atom travels relative to the crystal lattice. Moreover some of the displaced atoms recombine with vacancies in the lattice. Various estimates of the relationship between E_p and n_d, the number of atoms displaced, have been made. Figure 3.9 shows one such, that of Torrens, Robinson and Norgett (often called the "TRN", or alternatively "NRT", model), which is widely used.

It is possible to define a "displacement cross-section" for neutrons in group g, σ_{dg}, by

$$\sigma_{dg} \int_{E_g}^{E_{g-1}} \phi(E)dE = \int_{E_g}^{E_{g-1}} \sigma_e(E)n(E)\phi(E)dE, \qquad (3.12)$$

where

$$n(E) = \int_0^{\mu E} n_d(E_p)dE_p. \qquad (3.13)$$

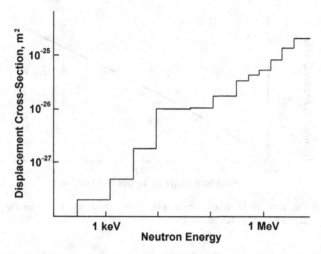

Figure 3.10 Group displacement cross-sections for iron (TRN model).

$\sigma_e(E)$ is the elastic scattering cross-section, and $\phi(E)$ gives the variation of flux within the group from, for example, a fundamental mode calculation. Figure 3.10 shows values of σ_{dg} for iron.

The total number of displacements suffered by each atom, D, is then given by

$$D = \int_0^T \sum_g \sigma_{dg}\phi_g(t)dt, \tag{3.14}$$

where T is the length of time for which the irradiation continues. D is the average number of times each atom is displaced from its lattice site and is often referred to as "dpa" (displacements per atom).

The scattering cross-sections of chromium and nickel are quite similar to that of iron so it is often assumed that the displacement cross-section of iron as shown in Figure 3.10 can be used for all the constituent elements of the steel structural materials used in reactor cores, whatever their actual specifications.

Figure 3.10 shows that σ_{dg} falls off rapidly at energies below about 0.1 MeV. For this reason an alternative way of characterising the extent of irradiation, often used because it is easier to calculate, is the fast neutron fluence, Φ_f, defined by

$$\Phi_f = \int_0^T \int_{0.1MeV}^\infty \phi(E,t)dEdt. \qquad (3.15)$$

The total flux above 0.1 Mev is sometimes called the "damage flux" and Φ_f is the "damage fluence". If the power density Q in mixed-oxide fuel containing 20% plutonium is 2.5 GW m^{-3} (see section 3.2.1) and the neutron spectrum is similar to that shown in Figure 1.7 the damage flux is about 5×10^{19} m^{-2} s^{-1}. If the fuel is irradiated to 20% burnup Φ_f is about 1.4×10^{27} m^{-2} and D is about 140 dpa. This irradiation takes about 5.4×10^7 s or 20 months at full power. These conditions are typical of the centre of the core and represent the full extent of the irradiation suffered by any of the structural material that is removed and replaced along with the fuel (i.e. cladding, subassembly wrappers, etc.). Any material in the core that is not replaced along with the fuel, such as the control rod guides, suffers a higher fluence. It is usual to design the core so that all of its structure can be replaced as necessary.

3.3.2 Irradiation Swelling

The cloud of vacancies and interstitial atoms produced by a neutron scattering event diffuses through the crystal lattice under the influence of thermal agitation. If the motion were entirely at random the density of vacancies and interstitials would rise until production was balanced by recombination at sinks such as grain boundaries and dislocations, where they would meet and annihilate each other. The motion is not entirely random, however. The stress field around a dislocation interacts with the stress fields around both interstitials and vacancies, and tends to attract them, but the interaction with an interstitial is stronger. As a result the interstitials tend to cluster together at dislocations and other defects in the crystals, leaving an excess of vacancies that also tend to form clusters rather than recombining. The normal form of

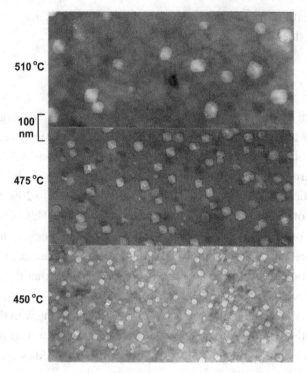

Figure 3.11 Electron micrographs of 20% cold-worked 316 stainless steel irradiated to 38 dpa (TRN).

a cluster of vacancies is a flat mono-atomic layer that eventually collapses leaving an edge-dislocation ring, but if there is a nucleus, which may consist of a few atoms of an inert gas (helium), vacancies migrate to it and form a three-dimensional void.

Figure 3.11 is a series of electron micrographs of irradiated AISI type 316 stainless steel showing typical polyhedral voids about 0.1 μm in diameter. As these voids are formed and grow the mean density of the material falls and it swells. Most metals swell in this way when irradiated but the rate and extent of swelling vary widely from one to another.

Temperature has an important effect on swelling as may be seen in Figure 3.11. Figure 3.12 shows the ranges of values of void

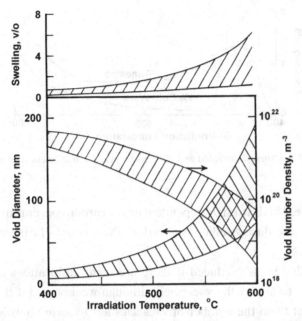

Figure 3.12 Void diameter and number density in cold-worked 316 stainless steel irradiated to 30–40 dpa (TRN).

density and mean diameter that are typically observed. The greater size of the voids at higher temperature is probably due to the increased mobility of vacancies at higher temperatures, but the corresponding reduction in the number of voids is not entirely understood because the mechanism of nucleation is uncertain. It may be that at high temperature the helium atoms migrate to existing voids and are not available as nuclei of new ones. Alternatively nucleation may be connected with the "spikes" of displaced atoms due to a neutron scattering event, the damage caused by which anneals out more readily at high temperature.

The result for many materials is that the swelling rate is high in a certain temperature range and low in others as shown in Figure 3.12. For 316 stainless steel there is little swelling below 350 °C. At high temperatures impurities may be very important, and above 600 °C the swelling is inhibited by the formation of very large voids on large grains

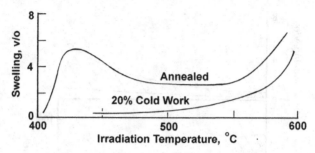

Figure 3.13 Swelling of annealed and 20% cold-worked 316 stainless steel at 35 dpa (TRN).

of carbide. This is strongly dependent on the carbon concentration, and one way to reduce swelling is to reduce the amount of carbon in the material.

Swelling is also reduced if there are many dislocations because they tend to attract the vacancies, although weakly, and if there are enough of them the number of vacancies left to form voids is small. This is shown by the much reduced swelling in 20% cold-worked 316 stainless steel as compared with annealed material, especially below 500 °C, as shown in Figure 3.13.

Finely dispersed precipitates within the grains also tend to reduce swelling because they attract both interstitials and vacancies and allow recombination. The nickel-rich alloy Nimonic PE16, in which the γ' phase is finely dispersed, is resistant to swelling for this reason. In a similar way the martensite crystals in ferritic-martensitic steels and the dispersions of nanometre-sized particles of oxide in oxide-disperse-strengthened or "ODS" steels have similar effects. As a result these materials may be particularly attractive for the cladding and the structure of the core.

In most materials the amount of swelling increases with time and because it depends on the interaction between vacancies and nuclei the increase is nonlinear. Initially swelling accelerates as irradiation proceeds, but at very high doses there is some evidence that it may saturate.

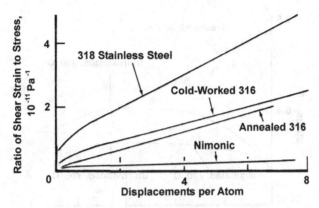

Figure 3.14 Irradiation creep in various materials.

3.3.3 Irradiation Creep, Embrittlement and Hardening

Under the influence of irradiation materials creep at low temperatures. The precise mechanisms of irradiation creep are not certain but in general terms it is clear that, as atoms displaced from their sites move about, they rearrange the crystal structure in such a way as to reduce the elastic energy, and if the material is under stress this gives rise to strain in the direction of the stress. For example one mechanism by which this might happen is that interstitial atoms produced by neutron scattering may tend to migrate to defects such as edge dislocations, causing them to climb so that the material strains. Another possible mechanism is that vacancies may coalesce on a plane in the crystal and if there is a compressive stress normal to the plane the resulting disc-shaped void may collapse causing the material to strain. At high temperature thermal agitation gives rise to such creep mechanisms: irradiation allows creep to take place at much lower temperatures. For example, substantial irradiation creep has been observed in 316 stainless steel at 280 °C whereas thermal creep is not significant below about 600 °C.

The creep strain is in most cases proportional to the stress, almost proportional to D, and nearly independent of temperature. Figure 3.14

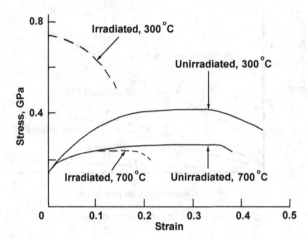

Figure 3.15 The effect of irradiation to 15 dpa on stress-strain curves for 316 stainless steel tested at 300 °C and 700 °C.

shows the dependence of the ratio of creep shear strain to stress on D for various materials.

Irradiation both hardens materials (i.e. it raises the yield stress) and makes them more brittle (i.e. it reduces the elongation before failure). The ultimate stress usually changes relatively little, but there is a loss of work hardening. Typical stress-strain curves are shown in Figure 3.15.

There are two mechanisms that cause these effects. At low test temperatures the defects caused by irradiation damage reduce the mobility of dislocations and inhibit plastic strain so that the uniform elongation is very low. At higher test temperatures (500–600 °C in 316 stainless steel), the material anneals, the defects are removed, and the properties of the unirradiated material tend to be recovered.

Above 700 °C in 316 stainless steel the second mechanism comes into play. As shown in Figure 3.15 there is a loss of ductility which can outweigh the increase in ductility in the unirradiated material due to thermal effects. This is the result of helium generated by (n, α) reactions. At high temperatures it diffuses to the grain boundaries

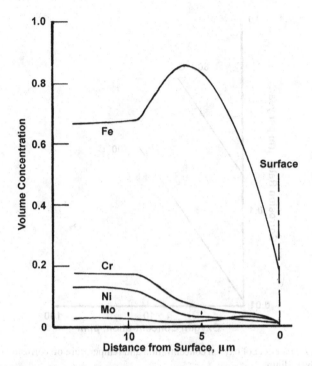

Figure 3.16 Depletion of Cr and Ni from the surface of stainless steel after exposure to sodium at 650 °C for 6 months.

where it collects in the form of small bubbles. These cause a loss of cohesion between the grains, so that they can be torn apart.

3.3.4 Corrosion in Sodium

Sodium has the effect of removing nickel and chromium from the surface layers of austenitic stainless steel. The chromium forms sodium chromite and the nickel is removed by dissolution in the sodium. Figure 3.16 shows the nickel and chromium concentrations near the surface of stainless steel exposed to flowing sodium. In this region the nickel concentration may be reduced to about 1% and chromium to 5–8%, and as a result a surface layer of ferrite is formed some

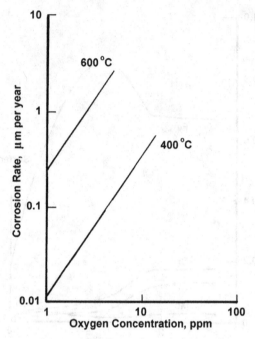

Figure 3.17 The effect of oxygen concentration and temperature on corrosion of stainless steel by sodium.

5 µm thick. This ferrite is then dissolved at a rate that depends on the oxygen concentration in the sodium, and as it dissolves the surface becomes rough. The greater the oxygen concentration, the greater the roughness. For 316 stainless steel, 10 parts per million of oxygen in the sodium give a roughness of about 2 µm, whereas 25 ppm give about 6 µm, more or less independent of the velocity or Reynolds number of the flowing sodium.

Because both the formation of the sodium chromite and the dissolution of the ferrite depend on oxygen, the corrosion rate is a function of the oxygen concentration as well as the temperature. Typical corrosion rates in 316 stainless steel are shown in Figure 3.17. The corrosion rates of low-alloy steel, such as 2.25 Cr 1 Mo, and also of PE16 alloy with its high iron content are very similar to those of stainless steel because in all cases the controlling process is the dissolution of ferrite. Figure 3.17 indicates that, with an oxygen concentration of around 10 ppm, which can be attained in practice in a fast power reactor, corrosion rates up to 10 μm per year are to be expected in the hottest parts of the coolant circuit where the temperature is around 600 °C. Higher oxygen concentrations carry a risk of excessive corrosion.

Corrosion or, more accurately, wear can be made worse by the relative motion of surfaces in contact, usually called "fretting". In a reducing environment such as sodium there is a tendency for surfaces in contact to become welded together at high spots. If the surfaces are forced to slide over each other, for example by some source of mechanical vibration, the welded high spots are sheared, and material may be transferred from one surface to the other. Fretting damage is a danger in parts of the core subject to vibration due to turbulent flow of the coolant. Self-welding is a problem for items of equipment such as fuel subassemblies that have to be removed from the reactor from time to time.

Depending on the activity of dissolved carbon, sodium can transfer carbon either to or from steels immersed in it. This process is known as carburization or decarburization depending on whether carbon is gained by the steel or lost from it. The rate of loss or gain of carbon depends on temperature because both the carbon activity in sodium and steel and the diffusion rates vary with temperature.

The carbon activity in austenitic steel is relatively low so it tends to be carburized even if the carbon concentration in the sodium is as low as 5 parts per million. Carbide is precipitated in the surface layers of the steel, reducing the ductility at low temperatures. Low-alloy ferritic steels, however, such as 2.25 Cr 1 Mo with about 0.1% carbon, tend to be decarburized and lose strength as a result.

Carbon activity is higher at low temperature, so there is a tendency for decarburization to be particularly important if a sodium circuit incorporates ferritic steels in a low-temperature region such as a heat exchanger. If excessive carburization of austenitic steels in the

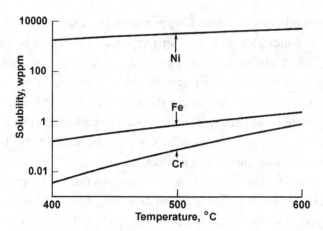

Figure 3.18 The solubility of iron, chromium and nickel in lead.

high-temperature regions is to be avoided, careful control of the carbon activity in the sodium has to be maintained.

3.3.5 Corrosion in Lead and Lead-Bismuth Eutectic

Iron, chromium and especially nickel are soluble in liquid lead. The solubilities are shown in Figure 3.18. In lead-bismuth eutectic (44.5Pb, 55.5Bi) the solubilities are about a factor of 10 higher. Unprotected steel surfaces corrode very rapidly by dissolution, not only of the metal surface but also beneath the surface as the liquid penetrates into defects and along grain boundaries.

The corrosion rate can be reduced substantially by a protective oxide layer on the surface of the steel. At temperatures in the range of 300–470 °C and oxygen concentrations of 10^{-9}–10^{-8} by atoms ($\sim 10^{-8}$–10^{-7}% by weight) an oxide layer a few tens of μm thick is formed that protects the steel from direct contact with the liquid metal. Corrosion continues by oxidation as oxygen diffuses through the layer. The layer grows in thickness until an equilibrium is reached, with oxide being formed at its inner surface and being lost by spalling from the outer surface. Figure 3.19 shows the composition of the oxide layer formed

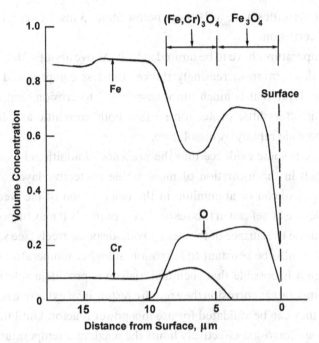

Figure 3.19 Formation of oxide layers on ferritic steel after exposure to lead-bismuth eutectic at 470 °C.

on a ferritic steel after exposure to flowing lead-bismuth eutectic at 470 °C. The outer part of the layer consists of porous Fe_3O_4 whereas the inner part is a compact $(Fe,Cr)_3O_4$ spinel. The protection is only partially effective, however, in that corrosion by oxidation continues at rates of 50 μm per year or more. Similar but thinner protective layers are formed on austenitic steels.

The steel surface is protected only while the oxide layer is in place. It can be disrupted mechanically by the impact of particles carried in the coolant, or even by the turbulence of the coolant itself. The pressure fluctuations in turbulent fluid flowing with velocity v are of the order of $\rho v^2/2$, and the high density of lead and lead-bismuth eutectic (for both of which $\rho \approx 10\,500\ \mathrm{kgm^{-3}}$) implies that the velocity has to be limited if the oxide layer is not to be eroded. In practice coolant velocities have to be kept below about 3 ms^{-1} to control the rate of corrosion.

Temperatures have to be limited as well. Above about 470 °C oxide is laid down in an increasingly thicker but less compact and more unstable layer that is much more susceptible to erosion, and above 550 °C it offers little protection so that both austenitic and ferritic steels corrode rapidly by dissolution.

There is some evidence that the presence of additional materials can result in the formation of more stable protective layer. A few percent of silicon or aluminium in the composition of the steel, for example, are beneficial in this respect. Alternatively it may be possible to aluminise the surface of the steel. Oxide-disperse steels (see section 3.3.8) may also be resistant to corrosion at higher temperatures. But although it is possible that such materials may provide a solution to high-temperature corrosion they require testing over extensive periods before they can be validated for use in a power reactor. Until this has been done corrosion effectively limits the maximum temperature in a reactor cooled by lead or led-bismuth eutectic to 470 °C.

3.3.6 Choice of Structural Materials

The advantages and disadvantages of the structural materials that can be used in the core can be summarised. There are three main groups: austenitic steels, nickel alloys and ferritic steels.

Austenitic steels of the AISI 304, 316 or 321 types are most widely used. They have relatively low yield strength but high ductility (which is particularly important for the fuel cladding). The creep rate can be reduced if the carbon content is reduced (as in the low-carbon 316 stainless steel) or if a stabilising additive such as niobium or titanium is included. In sodium systems these steels resist corrosion adequately provided the oxygen content is controlled. They suffer, however, from three disadvantages: irradiation embrittlement, susceptibility to damage by thermal shock, and, most important, irradiation swelling. Cold working to produce about 20% reduction of area reduces but does not eliminate the incidence of swelling (and also increases the yield stress), and irradiation creep mitigates the effects to some extent. Nevertheless the necessity to accommodate volume increases of 5–10% creates many difficulties in design.

High nickel alloys are considerably stronger and are much more resistant to irradiation swelling. They have the disadvantage that the neutron capture cross-section in nickel is higher than that in iron, but this is balanced by the higher strength, which allows a smaller volume of the material to be used. However alloys containing very high nickel levels (75% or so) corrode unacceptably fast in sodium. The high solubility of nickel rules nickel alloys out completely for lead or lead-bismuth systems.

Ferritic-martensitic steels suffer a loss of strength because of decarburization. This can be allowed for in design, and the effect is reduced if the carbon is stabilized by the addition of niobium or titanium. Ferritic steels have the significant advantages of a higher thermal conductivity and lower thermal expansion coefficient than austenitic steels, both of which reduce thermal stresses, and greater resistance to crack propagation. High chromium content such as in 9Cr 1Mo gives better resistance to corrosion in lead systems.

Ferritic steels lack creep strength at high temperatures and for higher temperature applications ODS steels may be preferable. Such steels are made by alloying the metal with fine oxide powders that after consolidation by hot extrusion form minute oxide crystals within the metal crystals. The particles consist of complex oxides of yttrium, titanium and aluminium, depending on the composition of the steel, and are typically 5 nm in diameter. They act to trap and immobilise dislocations, thus inhibiting creep, and also to attract vacancies, preventing them from coalescing to form voids and thus reducing swelling. ODS steels appear to be attractive as cladding and structural

Table 3.1 *Typical compositions of core structural materials (weight%)*

| Constituent | Austenitic | | Nimonic | Ferritic/Martensitic | | | ODS |
	316L	321	PE16	2.25Cr	9Cr	HT9	16CrODS
Cr	17.3	18.0	16.6	2.18	8.41	11.5	16.00
Ni	12.1	9.0	43.1	0.02	0.06	0.5	
Mo	2.3	0.5	3.4	0.92	0.88		
Mn	1.8	0.5	0.1	0.44	0.40	0.6	
Si	0.4		0.2	0.34	0.30	0.4	0.03
C		0.1	0.1	0.10	0.10		0.08
V				0.01	0.20		
W							1.82
Al			1.2				4.59
Ti		0.5	1.2	0.01	0.01		0.28
Y_2O_3							0.37
Fe	66.1	71.4	34.1	95.98	89.64	87.0	76.83

materials but they are difficult to weld and need extensive experimental validation before they can be used with confidence.

Table 3.1 gives typical compositions of some of these materials.

3.4 CORE STRUCTURE

3.4.1 Fuel Subassemblies

The design of fuel elements is discussed in Chapter 2. Each element, or
"fuel pin" as it is sometimes called, consists of a steel tube, 6 to 8 mm
in diameter and usually some 2.5 to 3 m long. It contains about 1 m or
more of core fuel, in the case of a breeder reactor 0.3 to 0.5 m of axial
breeder fuel both above and below the core fuel, and a void or plenum
up to 1 m long to contain the fission-product gases released from the
fuel as explained in section 2.3.6. The plenum can be situated either
below or above the core and axial breeders. A typical fuel element for
a breeder reactor is shown in Figure 2.6.

Unlike those of thermal reactors the fuel elements of a fast reactor are usually arranged in a triangular array. This is better than a square array because, for the same ratio of coolant and fuel volumes, it allows greater clearances between the fuel elements and therefore smaller variations of cladding temperature round each element.

As shown in section 3.2.2 a reactor core may contain some 10^5 fuel elements, which obviously cannot be inserted and removed one by one. They are assembled into subassemblies each consisting of 200–400 fuel elements surrounded by a hexagonal tube or "wrapper". In addition to facilitating handling this arrangement enables the coolant flow-rate to each part of the core to be regulated so that the outlet temperature is approximately uniform across the core. This is done by means of adjustable flow restrictors or "gags" at the bottom of each subassembly.

The use of subassembly wrappers underlines another important difference between fast and thermal reactors. In a thermal reactor the incorporation of large amounts of structural material in the reactor core would have an adverse effect on the neutron economy, but in a fast reactor in which the neutron cross-sections are much lower it has little effect. In addition the choice of the structural material is hardly constrained by neutronic considerations: there is no need to resort to zirconium; steel is entirely acceptable.

The choice of subassembly size is a matter of compromise. Large subassemblies make for easier fuel handling and lower manufacturing costs. Small subassemblies have the advantages of smaller transverse temperature variation (and therefore less severe hot spots), easier removal of decay heat when the irradiated fuel is removed from the reactor, and less serious consequences of an accident confined to one subassembly. (This point is explained more thoroughly in Chapter 5.)

There are two possible methods of locating the fuel elements within the wrapper. In one a wire is wrapped helically around each element with a pitch of the order of 0.2 m. The diameter of the wire is equal to the required distance between elements so that each element is located with respect to each of its six nearest neighbours at two points in each pitch of the helix. Wire wraps tend to induce helical distortion of the fuel elements because they cause a slight asymmetry of the cladding temperature, and they introduce a considerable quantity of metal into the core, but they have the advantage that the helical shape tends to promote mixing of the coolant between the various subchannels. In spite of this the edge subchannels adjacent to the wrapper are larger than the rest so the fuel elements at the edge of the bundle are slightly over-cooled (as shown in Figure 3.8).

An alternative to wire wraps is to locate the fuel element by means of transverse grids every 0.1 m or so along their length. Grids use less material and do not distort the elements as much as wire wraps, and they allow the possibility of reducing the over-cooling of the edge fuel elements. However they are expensive and have the disadvantage that the fine clearances between grids and elements may be prone to blockage by any foreign material present in the coolant. They also tend to have a high resistance to coolant flow, which increases the pressure drop through the core.

As well as fuel elements a subassembly may also contain shielding material, usually in the form of steel rods or a steel block, both above and below the fuel elements. The shield serves to protect the permanent structure on which the core rests, and equipment such as heat exchangers and the roof of the reactor vessel, from neutron and γ radiation from the core. There is also usually a coolant filter at the inlet end. Figure 3.20 shows a typical fuel subassembly. Radial breeder fuel is usually contained in larger diameter fuel elements in sub-assemblies similar to those in the core.

3.4.2 Subassembly Bowing and Restraint

The subassembly wrappers become distorted due to the effects of thermal expansion, irradiation swelling and irradiation creep. The resulting movements have to be checked; otherwise they may cause

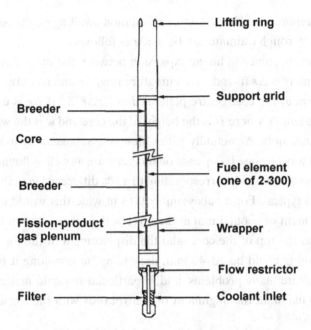

Figure 3.20 A typical fuel subassembly for a breeder reactor.

unwanted reactivity changes, interfere with movement of the control rods, and make withdrawal of irradiated subassemblies difficult.

As explained in section 3.2.5 there may be a considerable temperature difference across a subassembly at the side of the core due to the radial variation of power density. The side of the wrapper nearer the core centre is hotter and expands more, causing the subassembly to become curved or "bowed" with the convex side facing the core centre. The effect on reactivity of the resulting displacement, if it were unconstrained, is explained in section 1.6.3. A similar effect can arise from radiation-induced swelling of the wrapper, which is greater on the side nearer the core centre where the neutron flux is higher.

The effect of temperature bowing in an unconstrained core would in fact be small. Temperature differences of 10 K across subassemblies would cause radial displacements at the core centre plane of the order of 0.2 mm and reactivity changes of the order of 10^{-4}, which are of minor importance. The effect of irradiation swelling could be much greater. A rough estimate can be made as follows.

If the difference in linear expansion between the two sides of the subassembly is $\Delta\varepsilon$ its radius of curvature is $w/\Delta\varepsilon$ and its outward displacement at the core centre plane is $d \approx H^2\Delta\varepsilon/8w$ if $\Delta\varepsilon$ is uniform along its length, where H is the height of the core and w is the width of the subassembly. $\Delta\varepsilon$ actually varies along the subassembly in a complicated way because it depends on temperature as well as fluence, but a mean value of 0.003, corresponding to a 1% difference in volumetric strain, is typical. For a subassembly 0.15 m wide this would cause a displacement of about 3 mm at the level of the core centre and about 10 mm at the top of the core, and the displacement of the top of the subassembly could be 20–40 mm, depending on how long it is. This would create many problems, and in particular it would make it difficult to maintain the alignment of control rods with their operating mechanisms.

In most reactors radial movement is prevented by some sort of restraining or clamping system. Figure 3.21 shows such a system, where radial movement of the outer periphery of the radial breeder is prevented at two restraint planes, both above the core. According to where the restraints are placed small movements of the fuel due both to thermal expansion and to irradiation swelling still occur, but because the subassemblies are in contact with each other across the core any tendency to positive reactivity feedback can be eliminated.

A restraint system may be "active", meaning that after the subassemblies have been assembled the restraints are tightened so that the whole array is clamped together, or "passive", meaning that the restraints are fixed and prevent outward movements only after clearances between subassemblies have been taken up as the wrappers swell. The disadvantage of an active system is that the clamping mechanism has to be operated remotely and reliably under sodium.

To avoid self-welding between the wrappers they can be provided with hard pads at the points where they are in contact. This is one

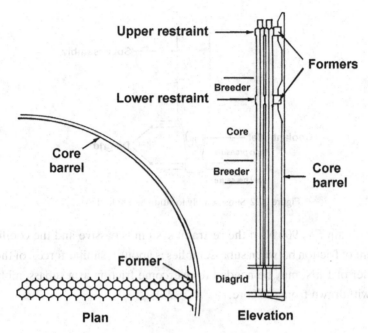

Figure 3.21 Core restraint.

of the few places where neutronics affects the choice of materials: the cobalt alloys that are usually used for hard-faced surfaces are not acceptable in a reactor core because the radioactive ^{60}Co that would be generated would create severe problems in handling the irradiated fuel and disposing of radioactive waste.

The radial loads between the restrained subassemblies vary with time and depend on the design of the restraint system, and predicting them is a very complex task. The order of magnitude can be estimated fairly simply, however. If a simple cantilever of length L is subjected to a transverse force F at its free end, the displacement d is given by $d = FL^3/3EI$, where I is the second moment of area of the cross-section and E is Young's modulus. For a hexagonal wrapper 0.14 m across flats made of sheet 3 mm thick, $I \approx 4 \times 10^{-6}$ m^4. If the length of the subassembly, L, is 3 m and the displacement d is 0.01 m, typical of bowing due to irradiation swelling, then with $E = 2 \times 10^{11}$ Pa

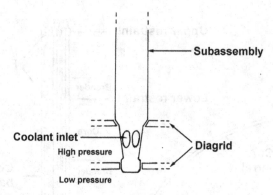

Figure 3.22 Subassembly location and hold-down.

we obtain $F \approx 900$ N. If the restraint system is passive and the coefficient of friction between subassemblies is about 1, similar forces, of the order of 1 kN, may be needed to overcome friction as a subassembly is withdrawn from the core.

3.4.3 Diagrid

The subassemblies rest on a support structure, often known as a "diagrid", which also serves to distribute the coolant from the circulating pumps. A typical arrangement is shown in Figure 3.22: each subassembly terminates in a hollow spike that is located in a hole in the diagrid, and coolant enters the subassembly through slots in the spike. If the coolant is a liquid metal, when it is flowing the pressure inside the diagrid is some 200–400 kPa above the hydrostatic pressure of the coolant outside, and this causes a substantial lifting force on a subassembly, which can exceed its weight. To prevent it floating away a "hold-down" mechanism is required. In the case of sodium coolant this can be done by arranging for the spike to pass right through the diagrid, so that its lower end is exposed to coolant at low pressure.

Alternatively, for sodium, hold-down can be provided by a rigid structure above the core that prevents upward movement. In the case of lead coolant a hold-down structure above the core is

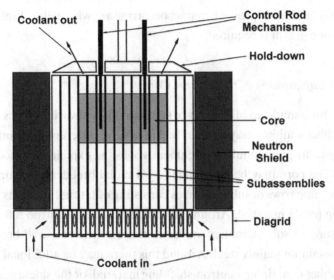

Figure 3.23 Diagrid, subassemblies and hold-down in a power reactor.

essential because otherwise the subassemblies would float away when the coolant flow-rate and the pressure in the diagrid are low. The hold-down structure has to make provision for the coolant flow to leave the subassemblies and for the control-rod drive mechanisms to operated through it, and also to allow for the subassemblies to be removed and replaced. Figure 3.23 shows, in diagrammatic form, how diagrid, subassemblies and hold-down are commonly arranged.

The wrappers contain high-pressure coolant from the diagrid but the space between the wrappers is open at the top to low-pressure coolant, so that there is a pressure difference across the wrappers of the order of 100 kPa. This induces stresses, and in the core region, where the neutron flux is significant so that irradiation creep takes place, the flat faces of the hexagonal wrappers bulge outwards. If this brings neighbouring wrappers into contact self-welding may occur and withdrawal of irradiated subassemblies may be difficult. The size of the gap between wrappers, which is determined by the thickness of the hard pads where contact between subassemblies is allowed, must be great enough to prevent contact occurring anywhere else. Typically a gap of 6–8 mm is required.

3.4.4 Configuration of the Reactor Core

A reactor is made up of an array of subassemblies of various types. The core subassemblies may contain fuel of several different enrichments arranged to give annular enrichment zones, as explained in section 1.3.3. The core may be surrounded by a radial breeder consisting of two or three rows of subassemblies consisting of fat fuel elements containing fertile material. Around this there may be a neutron reflector consisting of subassemblies similar to those of the core and breeder but containing mainly steel. Around this there may be additional subassemblies containing neutron-shielding material, or the shield may be a separate structure as indicated in Figure 3.23. Control rods occupy subassembly positions in the core region and are usually inserted into the core from above. They are operated by mechanisms that are situated on top of the reactor vessel so that they are available for maintenance.

Figure 3.24 shows a core pattern for a 3000 MW(heat) sodium-cooled breeder reactor. The core has two fuel enrichment zones and is surrounded by a radial breeder, which in turn is surrounded by a reflector. The neutron shielding is not shown. The effective diameter of the core is 3.6 m and it is 1 m high. Figure 3.25 shows a much smaller 600 MW(heat) sodium-cooled core that is 1.5 m in diameter and 0.9 m high. As explained in section 3.2.3 the height of the core of a sodium-cooled reactor is constrained by coolant flow considerations to be about 1 m, and the core diameter is adjusted to accommodate the required power output.

The configuration is different if the reactor is intended to consume rather than breed fissile material. Figure 3.26 shows how the core shown in Figure 3.24 could be modified for this purpose. There is no

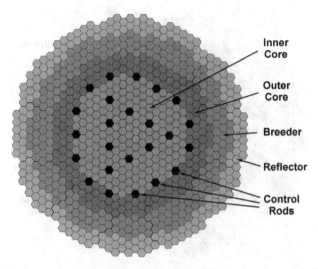

Figure 3.24 The configuration of the core of a 3000 MW (heat) sodium-cooled breeder reactor.

breeder, and in addition some of the ^{238}U is removed from the core (by increasing the fuel enrichment). This would make the core excessively reactive so some of the fuel subassemblies have to be removed and replaced by diluent subassemblies containing inert material.

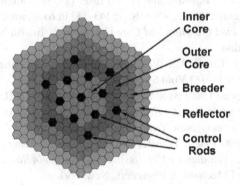

Figure 3.25 The configuration of the core of a 600 MW (heat) sodium-cooled breeder reactor.

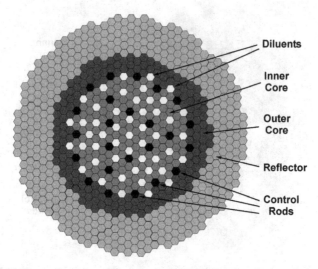

Figure 3.26 The configuration of the core of a 3000 MW (heat) sodium-cooled consumer reactor.

REFERENCES FOR CHAPTER 3

Allen, T. R. and D. C. Crawford (2007) Lead-Cooled Fast Reactor Systems and the Fuels and Materials Challenges, *Science and Technology of Nuclear Installations*, article ID 97486

Bagley, K. Q., J. W. Barnaby and A. S. Fraser (1973) Irradiation Embrittlement of Austenitic Stainless Steels, pp 143–153 in *Irradiation Embrittlement and Creep in Fuel Cladding and Core Components*, British Nuclear Energy Society, London

Bramman, J. I., C. Brown, J. S. Watkin, C. Cawthorne, E. J. Fulton, P. J. Burton and E. A. Little (1978) Void Swelling and Microstructural Changes in Fuel Pin Cladding and Unstressed Specimens irradiated in DFR, pp 479–508 in *Radiation Effects in Breeder Reactor Structural Materials*, American Society of Mining Engineers, New York

Dwyer, O. E. (1968) Heat Transfer to Liquid Metals flowing In-line through Unbaffled Rod Bundles, pp 139–168 in *Heat Transfer in Rod Bundles*, American Society of Mechanical Engineers, New York

Etherington, E. W., J. I. Bramman, R. S. Nelson and M. J. Norgett (1975) A UKAEA Evaluation of Displacement Damage Models for Iron, *Nuclear Engineering and Design*, 33 82–90

Friedland, A. J. and C. F. Bonilla (1961) Analytical Study of Heat Transfer Rates for Parallel Flow of Liquid Metals through Tube Bundles, *Journal of the American Institute of Chemical Engineering*, 7, 107–112

Hoffman, H. and D. Weinberg (1978) Thermodynamic and Fluiddynamic Aspects in Optimizing the Design of Fast Reactor Subassemblies, pp 133–139 in *Optimisation of Sodium-Cooled Fast Reactors*, British Nuclear Energy Society, London

Hsiung, L., M. Fluss and A. Kimura (2010) *Structure of Oxide Nanoparticles in Fe-16Cr MA/ODS Ferritic Steel* Lawrence Livermore National Laboratory report LLNL-JRNL-427350

Mosedale, D. and G. W. Lewthwaite (1974) Irradiation Creep in Some Austenitic Stainless Steels, Nimonic PE16 Alloy, and Nickel, pp 169–188 in *Creep Strength in Steel and High-Temperature Alloys*, London, The Metals Society, London

Nettley, P. T., I. P. Bell, K. Q. Bagley, D. R. Harries, A. W. Thorley and C. Tyzack (1967) Problems in the Selection and Utilization of Materials in Sodium Cooled Fast Reactors, pp 825–849 in *Fast Breeder Reactors* (BNES Conference proceedings), Pergamon, Oxford

Subbotin, V. I., A. K. Papovyants, P. L. Kirillov and N. N. Ivanovskii (1963) A Study of Heat Transfer to Molten Sodium in Tubes, *Soviet Journal of Atomic Energy*, 13, 991–994

Tang, Y. S., R. D. Coffield and R. A. Markley (1978) *Thermal Analysis of Liquid-Metal Fast Reactors*, American Nuclear Society, Hinsdale, Illinois, USA

Thorley, A. W. and C. Tyzack (1973) Corrosion and Mass Transport of Steel and Nickel Alloys in Sodium Systems, pp 257–273 in *Liquid Alkali Metals*, British Nuclear Energy Society, London

Zhang, J. and N. Li (2007) Review of the Studies on Fundamental Issues in LBE corrosion, *Journal of Nuclear Materials*, 373, 351–377

4

COOLANT CIRCUITS AND
STEAM PLANT

4.1 INTRODUCTION

4.1.1 *Choice of Coolant*

This chapter describes the engineering of the remainder of the plant
in a fast reactor electricity-generating station, apart from the reactor
core that is the subject of Chapter 3. The nature of the plant depends
primarily on the coolant, which is the heat-transfer medium. The main
considerations determining the choice of the coolant were explained
in sections 3.2.3 and 3.2.4. The most important is that the high power
density of a fast reactor core demands a high-density coolant and high
coolant velocities. The relative advantages and disadvantages of the
various possible coolants can be summarised in terms of the choices
available to a reactor designer, as follows.

Liquid or Gas. Helium has the advantage that it is chemically inert
and is therefore appropriate for use in a high-temperature reactor.
CO_2 has the advantage that there is extensive experience of its use in
thermal reactors. Neither presents significant problems of corrosion
or erosion. However any gas coolant has to be pressurised to make
it dense enough to transport heat out of the core without unreas-
onably high velocities. The major consequent disadvantage is that it
is then very hard to guarantee that decay heat could be removed
safely in the event of an accidental loss of pressure. It would be

necessary either to accept relatively low power density in the core (compared with what is possible if a liquid coolant is used) or to provide elaborate emergency cooling equipment for use in the event of a major breach of the primary coolant system. For this reason no gas-cooled power-producing fast reactor has, at the time of writing, been built and thus there is no operating experience, but that does not mean that gas coolant may not at some time in the future become attractive.

Water or Liquid Metal. Water is almost inevitably ruled out as a coolant for a fast reactor because the moderating effect of the hydrogen would degrade the neutron energy spectrum to the extent that its advantages – either as a breeder of fissile material or as a consumer of radioactive waste – would be lost or at least drastically reduced. A fast reactor cooled with supercritical water ("supercritical" in the thermodynamic sense, at a pressure above the critical pressure of 22.12 MPa) has been suggested but never taken beyond the stage of an outline design.

All liquid metals have the major advantage that they do not have to be pressurised so the reactor structure can be relatively light. In the case of an accident the decay heat can be removed by ensuring that the fuel stays immersed in the coolant, and it may be possible to arrange the primary coolant circuit so that even if the pumps fail natural convection cooling is adequate.

An important disadvantage of liquid metals is that they are opaque, which makes inspection of the core and coolant circuit structures and components difficult.

Light or Heavy Liquid Metal. The alkali metals lithium, sodium and potassium all suffer from the major disadvantage that they react chemically with air and water. They have the advantage that they are light, and they are not corrosive. They have low melting temperatures so that it is relatively easy to avoid freezing, but also low boiling temperatures so that there is a possibility that the cooling may be impaired under extreme accident conditions. They are all moderators.

Sodium and potassium are cheap. Lithium is too expensive and too much of a moderator to be considered.

Heavy liquid metals such as lead or bismuth have the major disadvantages that they corrode steel and that at more than moderate velocities they cause erosion and cavitation damage, particularly in pump rotors. In addition they are heavy and expensive and they have high melting temperatures. Their advantages are that they do not react chemically with water, they have high boiling temperatures, and they are poor moderators (so they do not degrade the neutron energy) while having high scattering cross-sections (so they reduce neutron leakage from the core).

Mercury was the coolant for very early experimental fast reactors in the United States and the former Soviet Union but has never been used since. It would not be contemplated now because it is too expensive and too toxic.

Sodium or Potassium. Although both are chemically reactive potassium is rather more hazardous than sodium. Compared with potassium, sodium has the disadvantage that it becomes radioactive by the $^{23}Na(n,\gamma)^{24}Na$ reaction. The resulting ^{24}Na decays with a 15-hour half-life. While the reactor is operating the specific activity of the sodium primary coolant may exceed 30 GBq/kg.

It has never proved possible to eliminate the possibility of a leak in a sodium-heated steam generator, and in fact such leaks have occurred quite frequently. A large steam-generator leak generates large volumes of steam and hydrogen accompanied by tens or hundreds of kilograms of NaOH, and the only way to protect the reactor is to vent these reaction products to the atmosphere. This cannot be contemplated if they are radioactive. Therefore the steam generators cannot be heated directly by the radioactive primary sodium, and intermediate nonradioactive secondary sodium circuits have to be interposed.

Sodium, being lighter, has a greater moderating effect than potassium. This is a disadvantage not only in that it degrades the neutron energies but also because removal of sodium from the centre of the core causes a positive reactivity change. This gives an undesirable positive contribution to the temperature coefficient of reactivity and, in a severe accident that causes the sodium to boil, a substantial reactivity increase.

However sodium has the overwhelming advantage that it is cheap and readily available, and for this reason, in spite of its disadvantages, it has been used almost universally as the coolant for fast power reactors.

Pure potassium has never been used as a reactor coolant but in a few cases sodium-potassium alloy, usually referred to as "NaK", has proved attractive. Pure sodium melts at 97.8 °C so care has to be taken to avoid freezing, but the admixture of potassium reduces the melting point. Eutectic NaK (77%K, 23%Na) freezes at −12.6 °C.

Lead or Lead-Bismuth Alloy. The melting point of pure lead is 327 °C so extensive reliable trace heating of the primary coolant circuit has to be employed to avoid freezing. However, as in the case of NaK, the addition of bismuth can reduce the requirement substantially. Eutectic Pb-Bi (55%Bi, 45%Pb) freezes at 123.5 °C.

The main disadvantage of lead-bismuth alloy is the reaction $^{209}Bi(n,\gamma)^{210}Bi$. The resulting ^{210}Bi is β-active with a half-life of 3.5×10^6 years, producing ^{210}Po which is α-active with a half-life of 138 days and is a major hazard in reactor maintenance and refuelling. In addition bismuth is expensive. In spite of these disadvantages lead-bismuth was chosen as the coolant for the fast reactor power plants of the Soviet "Alpha" class submarines, and their deployment has provided many reactor-years of operating experience. When in the 1990s the Russian authorities released this information lead-bismuth became a serious alternative to sodium as a coolant for fast reactors.

4.1.2 Sodium Coolant

EBR-1, which started operation in the United States in 1951, was a small fast reactor cooled with sodium. Its operation was largely successful and was widely reported. Possibly for these reasons it set a trend for sodium cooling, and designers of subsequent civil fast reactors kept to a coolant that was known to work rather than risk trying something else. As a result almost all fast reactors have been sodium-cooled. The only exceptions have been the USSR submarine reactors mentioned earlier for which sodium could not be used because of the impossibility of venting sodium-water reaction products.

Because there is such a preponderance of sodium design and operating experience, and because at the time of writing all operating and planned fast reactors are sodium-cooled, the rest of this chapter deals only with the sodium systems of power-producing fast reactors. It concentrates on the primary and secondary coolant circuits. The associated steam plant, apart from the steam generators, is very similar to that in a conventional fossil-fuelled power station so it is mentioned only briefly here, and the electrical equipment, which is entirely conventional, is not dealt with at all.

Chapter 3 emphasises the way in which the designer of the reactor core is constrained within very narrow limits by the properties of the materials so that there is a marked similarity between all fast reactor cores. The same is not true for the coolant circuits. The use of sodium dictates the size of the heat transfer surfaces but not the form of the circuits or the heat exchangers. There are for example two very different approaches to the layout of the primary circuit, leading to either a "loop" or a "pool" reactor. Similarly many different steam generator designs are possible, employing U-tubes, straight tubes or helical tubes, with either once-through steam flow or separate evaporators and superheaters.

So far all fast power reactors have had a secondary liquid metal coolant circuit because the risk of having water and radioactive primary sodium in the same heat exchanger has been judged unacceptable. With increasing experience of designing and operating sodium-heated steam generators, and of preventing leaks in them, the situation may change in the future, because the capital cost would be reduced if the secondary sodium circuits could be dispensed with. But because the present purpose is to describe existing and imminent systems only designs incorporating secondary sodium circuits are considered here.

4.2 PRIMARY SODIUM CIRCUIT

4.2.1 Pool or Loop Layout

The primary coolant receives heat in the reactor core, flows to a region where the neutron flux is low to transfer its heat to the secondary coolant in the intermediate heat exchangers, and then returns to the core. In fixing the layout of this primary coolant circuit two main choices have to be made: whether the heat exchangers and circulating pumps should be in separate vessels from the core or in the same one, and whether the pumps should be located before or after the heat exchangers.

A "pool" reactor is one in which the entire primary circuit is contained within a single vessel, as shown in Figure 4.1 A. The core is surrounded by a neutron shield and around this are placed the pumps and heat exchangers. In a "loop" reactor, in the other hand, as shown in Figure 4.1 B, the core is contained in a small vessel with the main neutron shield outside. Hot coolant from the core passes through pipes to the heat exchangers and then back to the core vessel.

The choice between the two schemes is affected by such considerations as the design and manufacture of the vessels, the design of the refuelling system, the operating conditions of the pumps, and ease of inspection and maintenance. That the choice is finely balanced is shown by the fact that reactors of both types have been built, but most proposed future reactors are of the pool type.

The main advantages of the pool layout are that the coolant pressure-drop is low and the reactor vessel is very simple in shape without irregularities that might act as locations of high stress. The primary circuit can be arranged so that hot coolant never comes into

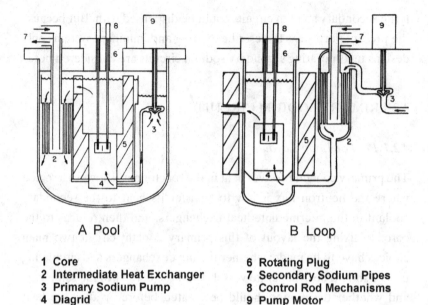

A Pool   B Loop

1 Core   6 Rotating Plug
2 Intermediate Heat Exchanger   7 Secondary Sodium Pipes
3 Primary Sodium Pump   8 Control Rod Mechanisms
4 Diagrid   9 Pump Motor
5 Neutron Shield

Figure 4.1 Pool and loop layout of the primary circuit.

contact with the vessel. In a loop reactor parts of the pipework and of the vessel are in contact with hot coolant at temperatures at which thermal creep may be important, while other parts are in contact with cold coolant so that thermal stresses have to be allowed for. In addition there may be the possibility of stress concentrations caused by the pipe branches on the sides of the vessel.

On the other hand a pool reactor vessel is so large that it has to be assembled on site whereas a loop reactor vessel can be made in a factory where the quality of manufacture can be controlled more easily. The roof of a pool reactor vessel is much larger than that of a loop vessel and if advantage is to be taken of the potential simplicity of the vessel itself the pumps and heat exchangers, and possibly the entire core and neutron shield, have to be suspended from it. Moreover part of the underside of the roof is exposed to the temperature of the hot coolant. As a result it is a complex and expensive structure.

A pool reactor has the advantage that there is room within the vessel for a temporary store for irradiated fuel, usually surrounding the neutron shield. Fuel can be transferred from the core to the store without lifting it above the coolant so that no special provision has to be made for cooling it while in transit. It can be left in the store, immersed in coolant, until the fission-product decay power has decayed sufficiently to make handling easier when it is removed for reprocessing. A loop reactor vessel is unlikely to have enough room for an irradiated fuel store. Irradiated fuel has to be removed from the vessel to a separate store by a machine that is capable of cooling it while it is in transit.

Both pool and loop reactors have pipework or structure operating at the temperature of the hot coolant. The difference is that in a loop reactor the hot pipework is part of the primary coolant containment, and if it should fail radioactive primary coolant could be released. To offset this disadvantage, however, the loop system has the advantage that it may be possible to inspect the high-temperature pipework more easily because it is accessible from outside. It may even be possible to do maintenance work on one coolant loop by closing it off with valves without shutting down the whole reactor. Inspection and maintenance of the pipework and structure within a pool reactor vessel are difficult.

Figures 4.2 and 4.3 show a typical arrangement of the components of a pool reactor. In the arrangement shown the core and neutron shield are supported by a strongback attached to the bottom of the vessel while the other main components – the primary pumps and the intermediate heat exchangers – are supported by the roof of the vessel. An alternative arrangement is to hang the core and shield from the roof as well. This has the advantages that the vessel carries only the weight of the sodium it contains and being simply shaped is relatively lightly stressed, and that stresses due to thermal expansion are minimised. Another alternative is to support the vessel and the core from below, but in this case thermal expansion stresses are greater.

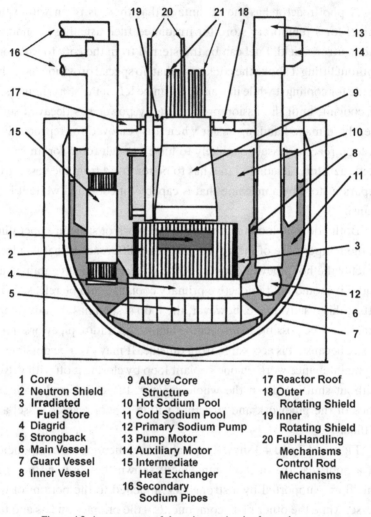

1 Core	9 Above-Core	17 Reactor Roof
2 Neutron Shield	Structure	18 Outer
3 Irradiated	10 Hot Sodium Pool	Rotating Shield
Fuel Store	11 Cold Sodium Pool	19 Inner
4 Diagrid	12 Primary Sodium Pump	Rotating Shield
5 Strongback	13 Pump Motor	20 Fuel-Handling
6 Main Vessel	14 Auxiliary Motor	Mechanisms
7 Guard Vessel	15 Intermediate	Control Rod
8 Inner Vessel	Heat Exchanger	Mechanisms
	16 Secondary	
	Sodium Pipes	

Figure 4.2 Arrangement of the primary circuit of a pool reactor.

A typical pool reactor vessel may be about 17 m in diameter and 16 m deep, containing about 2000 tonnes of primary sodium and made of stainless steel about 20 mm thick. It is surrounded by a second "leak jacket" or "guard vessel", so that even if the main vessel should break the sodium level cannot fall below the top of the core and emergency

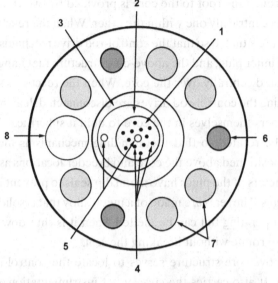

1 Above-Core Structure   5 Fuel-Handling Mechanisms
2 Inner Rotating Shield   6 Primary Sodium Pumps (3)
3 Outer Rotating Shield   7 Intermediate Heat Exchangers (6)
4 Control Rod Mechanisms   8 Fuel Transfer Position

Figure 4.3 Plan of the roof of a pool reactor with three secondary circuits.

cooling can be maintained. The space between the vessels provides access for inspection of the main vessel, for example by means of a remotely-controlled vehicle carrying an ultrasonic probe that can examine the welds, particularly those where the stongback is attached.

The inner vessel separates the hot and cold parts of the primary coolant circuit so that all the sodium in contact with the main vessel is at the cooler core inlet temperature. In some cases the inner vessel has a double wall to minimise the transfer of heat between the hot and cold sodium, but an alternative is to shape the lower part of the inner pool so that the sodium in it remains stagnant and acts as an insulating layer.

The reactor roof is a complex structure. Part of its underside faces the surface of the hot pool and has to be protected by thermal insulation, usually in the form of thin steel plates, and by a cooling system.

Access through the roof to the core is provided by two rotating plugs mounted eccentrically one within the other. When the reactor is operating they are situated so that the control-rod drive mechanisms mounted on the inner plug, and the above-core structure that hangs beneath it, are located centrally over the core. When the reactor is shut down for refuelling the control-rod drives are disconnected, leaving the control absorbers themselves in the core to hold it subcritical. The plugs can then be rotated so that fuel-handling mechanisms mounted on them are positioned above the core and breeder locations as required. The perimeters of the plugs have metal dip seals to prevent leakage of the cover gas. The seals are made of a metal alloy that is solid while the reactor is operating but can be melted when it is shut down to allow the plugs to rotate without breaking the seal.

The above-core structure serves to locate the control-rod drives accurately. It also carries the core outlet instrumentation, consisting of thermocouples to measure the coolant outlet temperature from each subassembly, and in many cases also coolant sampling take-offs that can be used to locate failed fuel (see section 5.2.3).

The vessel for a loop reactor may have inlet and outlet primary coolant connections in the side walls. Some loop reactor designs, however, use a larger vessel so that there is room for the inlet and outlet pipes to pass through the roof and the vessel itself can be simple in form and lightly stressed like a pool reactor vessel. This gives the additional advantage that more space for fuel handling or storage within the vessel may be made available.

The advantages and disadvantages of loop and pool schemes are discussed in detail by Campbell (1973).

4.2.2 Pumps

Some early fast reactors used electromagnetic pumps to circulate the coolant, which have the advantage that no moving part penetrates the sodium containment. The sodium is pumped either by passing an electric current through it in the presence of a transverse magnetic field (a conduction pump) or by subjecting it to a moving magnetic field (an induction pump). However it proved difficult to scale electromagnetic pumps up to the size needed for large reactors and now mechanical pumps are used universally. The problem of penetrating the sodium containment can be met by means of electric motors with totally enclosed, "canned", rotors, but the usual method is to allow the shaft to pass through the containment above the sodium level. The penetration is thus exposed to the argon cover gas containing sodium vapour but not to liquid sodium, and oil seals have been found satisfactory in most cases.

The pumps are thus driven by motors situated on the roof of the reactor vessel (in the case of the primary pumps of a pool reactor) or the pump vessel (in the cases of a loop reactor and the secondary circuits of either style), via vertical shafts with thrust bearings at the top and sodium-lubricated sleeve bearings at the bottom. The long shafts of the primary pumps of a pool reactor are usually tubular with a large diameter to avoid whirling. The variable-speed motors are fitted with auxiliary or "pony" motors that are capable of turning the pump fast enough to maintain adequate flow to keep the fuel cool when the reactor is shut down. In the case of power failure these can be energised from a standby source such as a diesel generator to guarantee emergency cooling.

The pumps have to provide a large volume rate of flow at a relatively low pressure rise. A typical 3600 MW(heat) reactor would have three primary pumps each delivering about $8 \, \text{m}^3 \, \text{s}^{-1}$ at 500 kPa. Single-stage centrifugal impellers of conventional design are normally used. In most cases the primary pumps deliver the coolant vertically downwards and the pump volutes are designed so that the resulting axial thrust, which can be as much as 10 tonnes, is borne by the pump casing rather than the shaft.

The main difficulties in design are to cope with sudden changes in temperature and to prevent cavitation, and in these respects loop and pool reactors pose different demands. The coolant temperature has very little effect on cavitation, because even at 600 °C the saturation pressure is only 7 kPa. The important factor is the pressure at the pump inlet (the "net positive suction head"). The pressure of the cover gas above the sodium in the reactor or pump vessels is limited to some 1–200 kPa gauge to minimise leakage of radioactive material and sodium vapour. The pump inlet pressure is this gas pressure plus the hydrostatic head of the sodium above the pump. The risk of cavitation can be minimised by increasing the depth of immersion, decreasing the rotational speed (which implies increasing the rotor diameter) or shrouding the inlet to make the flow uniform. For the primary pumps of a pool reactor all these options imply increases in the dimensions of the primary vessel.

In a loop reactor, however, the suction pressure depends on whether the pump is located before or after the heat exchanger (i.e. whether it is in the "hot leg" or the "cold leg"). If it is before the heat exchanger there is a loss of pressure due to the flow in the pipe from the reactor vessel to the pump. If it is after the heat exchanger there is an additional loss of pressure due to the flow through the heat exchanger and more pipe. It is not easy to compensate for these pressure drops by positioning the pump at a lower level to increase the hydrostatic head, and anyway doing so would require still longer pipes. Thus the conclusion is usually reached that cavitation is avoided more easily if the pump is placed in the hot leg, before the heat exchanger. The relative advantages of hot leg and cold leg pumps are discussed by Campbell (1973).

Primary or secondary pumps may be exposed to sudden temperature changes in the event of emergency shutdowns ("trips") of either the reactor or the steam plant. This is not usually a problem in the case of the primary pumps of a pool reactor because they draw from the large mass (of the order of 2000 tonnes or more) of cold coolant filling the vessel. In a loop reactor, however, the situation is quite different. If the reactor is tripped a hot leg pump is subject to a rapid fall in temperature, and if a secondary heat exchanger is shut down because of a steam plant trip a cold leg pump is subject to a rapid rise. These temperature transients can be designed for, but they constitute a disadvantage of the loop layout.

4.2.3 Intermediate Heat Exchangers

The intermediate heat exchangers, in which heat is transferred from primary to secondary sodium coolant, are normally of a shell-and-tube design. Differential expansion of tubes and shell can be accommodated by means of expansion bellows or by bends in the tubes. To eliminate any possibility of radioactive primary coolant leaking into the secondary circuit the pressure of the secondary coolant in the heat exchanger has to be greater than that of the primary. The tubes are usually arranged in an annular bundle, with the secondary sodium flowing down through a central duct and then upwards through the tubes.

Because the coolants transfer heat so readily it is possible to keep the temperature difference between primary and secondary coolant small and yet keep the intermediate heat exchangers reasonably compact. For example with tubes 20 mm in diameter and coolant velocities of about 5 m s^{-1} a Nusselt number of about 10 on both shell and tube sides is possible (see section 3.2.4), giving surface heat transfer coefficients of about 3×10^4 W m^{-2} K^{-1}. If the wall thickness is 1 mm this gives an overall heat transfer coefficient U of about 10^4 W m^{-2} K^{-1}. The heat transfer rate Q, heat transfer area A, and the logarithmic mean temperature difference ΔT_m are related by

$$Q = UA\Delta T_m, \qquad (4.1)$$

so that if $Q = 3.6$ GW and $\Delta T_m = 30$ K an area of 12000 m^2 is needed for heat transfer. If the tubes are 8 m long and 20 mm in diameter some 24000 of them are needed, which could for example be arranged in six separate units each containing about 4000 tubes. The diameter of each of these tube bundles, allowing for the central secondary sodium inlet duct, would be about 2 m. The design of intermediate heat exchangers is discussed by Tang, Coffield, and Markley (1978), p. 319.

In a pool reactor the primary coolant is driven through the intermediate heat exchangers by the pressure difference due to the difference in levels between the hot coolant within the inner vessel and the cold coolant outside it.

In the event of an accident it might become impossible to reject heat from the secondary coolant or the steam plant, and an alternative means of removing the heat due to decay of fission products in the fuel would be needed (see section 5.2.4). For this reason an auxiliary secondary coolant system is provided. In a pool reactor this may take the form of separate auxiliary heat exchangers in the vessel, in which heat can be transferred to an emergency or "decay heat removal" cooling system. In a loop reactor there may be a separate auxiliary cooling loop in the primary circuit, or arrangements for emergency cooling of the secondary circuits.

4.2.4 Thermal Shock

A disadvantage of liquid metal coolants as compared with water or gas is that because of the high thermal conductivity temperature changes in the fluid are transferred readily to the structure, giving rise to thermal stresses. There are two areas of concern: thermal shock and high-cycle fatigue.

Thermal shocks arise from the sudden changes in plant operating conditions that happen when emergency action is taken, such as tripping the reactor or the steam plant. If the reactor is tripped the power falls very rapidly, more quickly than the primary pump speed can be reduced, causing the coolant temperature at the core outlet to fall rapidly. This may impose a severe thermal shock on the above-core structure, but in a pool reactor further effects are mild in comparison because mixing in the hot pool delays and smoothes out the temperature change so that the effect on the intermediate heat exchangers for example is attenuated. In a loop reactor, in contrast, significant thermal stresses may be induced in the coolant outlet nozzles from the primary vessel and in the intermediate heat exchangers.

The intermediate heat exchangers may be more at risk from a trip initiated in the steam plant. If for example the turbine stop valve is tripped shut and the steam system is blown down through the safety valves, heat transfer from the secondary sodium in the steam generators is reduced very quickly. The secondary sodium pumps have to be slowed down, ideally in step with the declining heat transfer, to keep the sodium temperature at the steam generator outlet constant. This is clearly very difficult, and any mismatch changes the temperature at inlet to the intermediate heat exchangers. It is important that the resulting transient stresses do not threaten their integrity.

As explained in section 5.3.3 the secondary circuits are likely to be equipped with fast-acting isolation valves to protect the heat exchangers from damage by the caustic reaction products from a steam generator leak. Operation of these valves is however likely to cause a rapid temperature increase in the heat exchangers because, especially in the case of a pool reactor, hot primary sodium continues to flow.

The reactor plant components, and in particular the intermediate heat exchangers, have to be designed to tolerate these thermal shocks. This can be done by using the methods of fracture mechanics. It can be assumed that there are crack-like defects of the maximum size that would escape detection during manufacture. The growth of these hypothetical defects in thermal shock incidents can then be predicted and shown to be negligible or to be so small that integrity is not threatened. However it may also be necessary to keep a record of the thermal shocks actually experienced throughout the life of the reactor to ascertain that the growth of the hypothetical defects is within the limits assumed in the design. If it is not – if for example there were to be an unexpectedly large number of reactor trips – it might be necessary to inspect the components at greatest risk such as the intermediate heat exchanger tube-to-tubeplate welds to assure that they have not been damaged.

4.2.5 High-Cycle Fatigue ("Thermal Striping")

If two turbulent streams of sodium at different temperatures meet and mix together the temperature in the resulting stream will fluctuate initially, although the fluctuations will be smoothed eventually by conduction. However if the sodium flows over a component of the structure the temperature fluctuations may be transferred to it and cause fluctuating thermal stresses. Over the lifetime of a reactor the number of such fluctuations may be very high indeed and there may be a possibility of high-cycle fatigue damage.

(In the context of sodium-cooled reactors the phenomenon of high-cycle thermal strain is sometimes known as "thermal striping". This name derives from a conceptual image of streams of coolant at different temperatures emerging from different fuel subassemblies of a reactor core. These are envisaged as giving rise to "stripes" of coolant at different temperatures twisting around each other in the region above the core and impinging on the structure. This concept is neither physically accurate nor helpful. It is more accurate to think of streams of coolant at different temperatures flowing side by side and mixing by the formation of eddies, as shown in Figure 4.4. The structure is then exposed at different times to hot eddies in an otherwise cold flow or cold eddies in a hot flow. The term "thermal striping" is misleading and should be avoided.)

The potential for damage can be illustrated simply as follows. If the surface of a fully restrained structural member is subjected to temperature fluctuations with a peak-to peak amplitude of ΔT the resulting thermal strain range is roughly $\alpha \Delta T$. For an austenitic steel with $\alpha = 18 \times 10^{-6}\,\mathrm{K}^{-1}$, if ΔT exceeds about 110 K, the strain range exceeds 0.002. If the temperature fluctuates with a frequency of order 1 Hz, then over a 40-year life at 80% load factor there will be 10^9

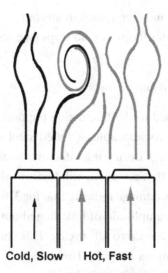

Cold, Slow   Hot, Fast

Figure 4.4 Flow mixing and eddy formation at the core-breeder boundary.

such cycles, which is enough to cause significant damage. If to avoid this damage the reactor is designed with a safety factor of 2 on strain, temperature fluctuations in excess of 55 K would have to be avoided. In a reactor where the core temperature rise may be 150 K or more this places severe restrictions on design, particularly for structures above the core outlet where temperature differences are greatest, both at the outlets of adjacent subassemblies and at the boundary of the core.

The component most likely to be at risk is the above-core structure. If for example there are two adjacent subassemblies from which coolant is emerging at different temperatures there will be temperature fluctuations in the region where the coolant streams mix. Figure 4.4 illustrates the situation. If the coolant velocity is $\sim 3 \text{ ms}^{-1}$ and the diameter of the subassembly outlets is ~ 0.1 m, eddies of ~ 0.03 m might be expected, with a frequency of ~ 1 Hz. The thermal diffusivity of sodium is about $6 \times 10^{-5} \text{ m}^2 \text{ s}^{-1}$, so the temperature difference in a typical eddy would last for 10 s or more, long enough for it to be transported anywhere in the hot pool and certainly for it to impinge on the above-core structure. Other components may also be at risk, not only in the main primary and secondary circuits but also in auxiliary circuits such as plugging meter loops or cold traps (see section 4.2.7).

4.2.6 Crack Initiation and Growth

If the surface exposed to the fluctuating temperatures is smooth and free from defects – for example, a rolled steel section that has not been subject to damage during manufacture or site assembly – it may be possible to show that no cracks will be formed. Experimental data on high-cycle fatigue damage indicate that for 316 stainless steel strain fluctuations with an amplitude of less than about 0.0008 do not initiate surface cracks even up to 10^9 cycles. This corresponds to surface temperature fluctuations of 45 K. This indicates that the above-core structure is not likely to be at risk if the differences between the coolant outlet temperatures from adjacent core or breeder subassemblies do not exceed this value.

In the case of a welded structure however it is very difficult to demonstrate freedom from surface defects 0.1–0.5 mm deep and it has to be recognised that temperature fluctuations may well make such cracks grow. In most cases they start to grow quickly, but the rate of growth soon declines. This is because high-frequency temperature fluctuations at the surface are attenuated rapidly as they penetrate into the material, and a growing crack soon reaches a depth at which they are too small to cause further growth. The crack growth is said to "arrest" at this depth.

It is usually the case that shallow cracks have no significant effect on the strength of reactor structures. Thus damage due to high-cycle fatigue can be taken to be acceptable if the crack arrest depth is 1 mm or less or, equivalently, if a crack of this depth will not grow. A crack grows if the variation of the stress intensity factor at its tip ΔK exceeds a critical value ΔK_{th}, the growth threshold. ΔK_{th} depends on the stress ratio R of the fluctuations. R is the ratio of maximum stress to minimum in the cycle, so that if the mean stress is zero $R = -1$,

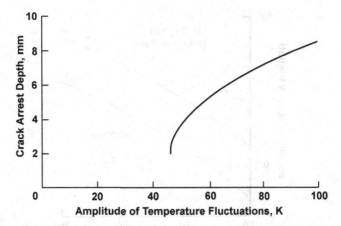

Figure 4.5 The dependence of the crack arrest depth on the amplitude of fluctuations in the surface temperature.

whereas $R = 0$ if the stress is zero at one extreme of the cycle. For 316 stainless steel at 550 °C ΔK_{th} is about 12 MPam$^{1/2}$ for $R = -1$ and 8 MPam$^{1/2}$ for $R = 0$.

A sinusoidal temperature fluctuation attenuates with depth x below the surface as $\exp(-x(\pi f/\lambda)^{1/2})$ where λ is the thermal diffusivity and f is the frequency, so at the tip of a crack of depth h the strain range $\Delta\varepsilon$ is given by

$$\Delta\varepsilon(h) = \alpha\Delta T_0 \exp(-h\sqrt{\pi f/\lambda}). \qquad (4.2)$$

The condition for crack arrest is then $h = h_c$, where

$$E\Delta\varepsilon(h_c)\sqrt{\pi h_c}/(1 - \nu) = \Delta K_{th}. \qquad (4.3)$$

Here E is Young's modulus ($= 2 \times 10^{11}$ Pa) and ν is Poisson's ratio ($= 0.3$). Figure 4.5 shows the variation of h_c with ΔT_0 for $f = 1$ Hz, $\alpha = 18 \times 10^{-6}$ K^{-1} and $\lambda = 6 \times 10^{-5}$ m^2 s^{-1}. For surface temperature fluctuations of $\Delta T_0 = 50$ K the crack arrest depth h_c is about 3.6 mm.

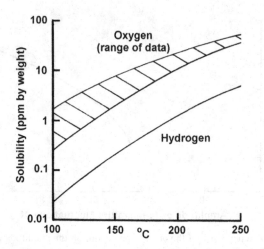

Figure 4.6 The solubility of oxygen and hydrogen in sodium.

4.2.7 Control of Impurities

The main impurities found in the sodium are oxygen and hydrogen. Oxygen enters the primary circuit as an impurity in the argon cover gas and as moisture on the surface of new fuel. Hydrogen, liberated by corrosion of the steam generator tubes, enters the secondary circuits by diffusing through the tube walls, and from there diffuses through the intermediated heat exchangers into the primary circuit.

As pointed out in section 3.3.4 oxygen dissolved in the sodium causes corrosion of stainless steel, and if the corrosion rate is to be limited the concentration of oxygen must not be too high. On the other hand there is some evidence that a little oxide in the sodium acts as a lubricant preventing self-welding between steel surfaces in contact and facilitating the operation of immersed mechanisms. It is usually controlled at a level of about 10 parts per million by weight or less.

The solubility of sodium monoxide Na_2O in sodium as a function of temperature is shown in Figure 4.6. If the oxygen concentration is high oxide tends to be precipitated in cool parts of the circuit. This is another reason for controlling the oxygen concentration because precipitation of oxide could block narrow passages such as coolant monitoring pipes.

Fortunately the low solubility of oxide at low temperature also affords means of controlling and monitoring the oxygen concentration. It can be controlled by passing the sodium through a "cold trap". This is a device in which the sodium is cooled to a temperature below 160 °C or so and then passed through a bed of stainless steel mesh or rings in which oxide is precipitated, so that the oxygen concentration is reduced to about 10 parts per million. To obtain lower oxygen concentrations a "hot trap" may be needed. In this the sodium is heated to 600 or 700 °C and passed over zirconium, which has a greater affinity than sodium for oxygen. A hot trap can reduce the oxygen concentration to 1–2 parts per million. Hot traps and cold traps cannot be used together because oxygen would be transported from the cold traps to the hot traps.

Hydrogen forms sodium hydride, NaH. The solubility of sodium hydride in sodium is also shown in Figure 4.6. It behaves similarly to oxide and is precipitated in cold traps in the same way. Because there is a continuous source of hydrogen cold traps in the secondary circuits have to be operated continually to prevent precipitation of hydride in the coldest parts of the steam generators. If it is not trapped in the secondary circuits hydride appears in the primary cold traps.

Tritium, ^{3}H, is radioactive with a half-life of 12.4 years, emitting a weak β. It is generated in the reactor core in three ways. It is formed directly in fission by rare ternary fission events, by the neutron capture reaction ^{6}Li (n,α) ^{3}H in lithium present as an impurity in the sodium, and by the capture reaction ^{10}B (n,2α) ^{3}H in boron in the control rods. The total production rate is of the order of 4 TBq per day (3×10^{16} atoms per second) in a 2500 MW (heat) reactor. Tritium diffuses readily through the tubes of the intermediate heat exchangers and tends to be precipitated in the secondary cold traps because these are usually in continuous operation. This may cause difficulty in disposing of the cold trap packing when it is full of hydride.

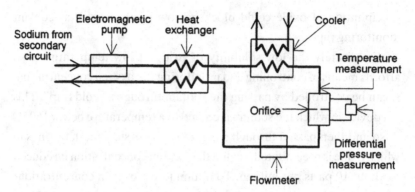

Figure 4.7 A plugging meter.

4.2.8 Monitoring of Impurities

The simplest and most effective means of monitoring impurity levels is a plugging meter, which measures the temperature at which solids are precipitated. Figure 4.7 shows the principle of a plugging meter. Sodium flows through an orifice, the pressure difference across which is monitored. As the sodium temperature is reduced the pressure difference rises when impurities are precipitated. A plugging meter is a valuable practical guide but gives little indication of what impurity is causing the blockage. It can tell the operator that an impurity is present, but not what it is.

Oxygen and hydrogen concentrations can be measured separately (Hans and Dumm, 1977). The oxygen concentration can be measured by an electrolytic meter in which the electrolyte is a ceramic (ThO_2 doped with YO_2) that separates the sodium, which forms one electrode, from an air reference electrode. The potential generated depends on the oxygen concentration in the sodium. The main problem is that the ceramic electrolyte is very brittle and susceptible to thermal shock, so the temperature of the sodium has to be controlled very carefully.

Hydrogen concentration can be measured by utilising its ability to diffuse through a nickel membrane into a carrier gas such as argon.

The concentration of hydrogen in the argon can then be measured by a katharometer, which depends for its operation on the marked effect of the hydrogen on the thermal conductivity of the mixture. Alternatively the membrane can be evacuated by a vacuum pump of known pumping speed. The hydrogen pressure upstream of the pump, which can be measured by an ionisation gauge, depends on the rate of diffusion through the membrane and therefore on the hydrogen concentration in the sodium.

4.2.9 Refuelling

A power reactor is usually designed to operate continuously for up to a year between refuelling shutdowns, at each of which up to half the core fuel, which has reached the end of its irradiation life, is removed and replaced by new.

New Fuel. New ceramic fuel subassemblies are usually delivered from the manufacturing plant in an atmosphere of air and can be kept in an air-filled store until required. Radioactive heating of new ceramic fuel is normally very slight so there is little requirement for cooling the store, but it has to be configured to prevent criticality. Before the fuel is committed to the reactor the store has to be purged with an appropriate gas. In the case of a gas-cooled reactor this would be the cooling gas, carbon dioxide or helium. For a metal-cooled reactor it would be the primary circuit cover gas – usually argon in the case of sodium coolant. The subassembly can then be transferred from the store to its required location in the reactor vessel where it is immersed in the primary coolant.

In the case of sodium coolant this is an irrevocable step, because once it has been wetted the subassembly cannot be withdrawn into an air environment until it has been thoroughly cleaned. This is because any residual sodium would react with atmospheric moisture to form caustic sodium hydroxide that might damage the cladding. If a new subassembly had for some reason to be withdrawn before it had been

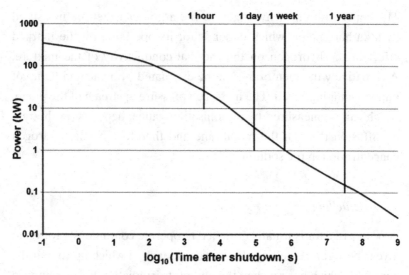

Figure 4.8 Subassembly decay heat after prolonged operation at 10 MW.

irradiated it would have to be inspected and requalified before it could be returned to use.

Recycled metal fuel in the IFR system (see section 2.5.6) contains fission products, the radioactive decay of which generates a significant amount of heat. New IFR fuel subassemblies have to be cooled continuously from manufacture until loaded into the reactor.

Irradiated Fuel. The route for withdrawing irradiated fuel from the reactor and dispatching it either for reprocessing or disposal as waste has to provide cooling to remove the decay heat produced by the fission products. It is important to prevent overheating which might cause failure of the cladding and release of radioactive fission products, fuel material or higher actinides to the environment.

The decay heat from a typical irradiated subassembly is shown in Figure 4.8. It depends slightly on the irradiation history, the fuel composition and the burnup. The high initial decay heat rating implies that an irradiated core subassembly has to be kept under forced-convection cooling for a period of around a day. In practice this means that the movement of irradiated fuel cannot start for a day after the reactor has been shut down. During the refuelling period the primary coolant is circulated, usually at around 10% of the full-power rate.

After this initial period the decay heat is low enough for the sub-assembly from a liquid-cooled reactor to be cooled by natural convection of the primary coolant. It can be removed from its position in the reactor core and placed in a storage position, but throughout the move it has to be kept immersed in the coolant. Only after a period of several months can it be withdrawn into a gas atmosphere.

The Fuel Transfer Route. The irradiated fuel store has to have the capacity to hold some 200 subassemblies. It can be located either within the primary vessel or in a separate vessel. In some older pool reactors it took the form of a rotating carousel in the primary vessel outside the neutron shield, but a more compact arrangement, which allows the main vessel to be smaller and therefore less costly, is to store the irradiated fuel in a ring around the core outside the neutron shield (see Figure 4.2). The alternative, which allows the main vessel to be even more compact, is to place the store in a separate vessel. The disadvantage of this is that it requires complex transfer equipment to lift the subassemblies out of the main vessel and lower them into the storage vessel while keeping them immersed in the coolant.

The transfer procedure for a sodium-cooled reactor is as follows. Fuel-handling machines are mounted on the eccentric rotating shields in the reactor roof (see Figure 4.3). The shields are manoeuvred to bring a handling machine over the subassembly to be moved. An arm is lowered from the machine and attached to the lifting ring at the top of the subassembly (see Figure 3.20). It then lifts the subassembly out of the core, keeping it below the surface of the sodium. The shields are then rotated to bring the subassembly over a transfer position into which it is lowered. A second move involving a second handling machine may be required to bring it to a position from which it can be removed from the reactor vessel.

At the removal position it is received into a cylindrical container, sometimes called a "bucket". The bucket full of sodium, with the subassembly in it, is then lifted up out of the reactor vessel and lowered down into the separate fuel storage vessel, where it is retained until it has cooled sufficiently to be removed from sodium. It is then lifted into an inert gas atmosphere, taken to a cleaning facility where the residual sodium is removed, and despatched to reprocessing or long-term storage.

4.3 STEAM PLANT

4.3.1 Steam Generator Design

The secondary sodium gives up its heat to raise steam in steam generators that are normally shell-and-tube heat exchangers with water or steam in the tubes. These have to be larger than the intermediate heat exchangers because of the poorer heat transfer on the steam side. They also differ in that they are stressed by the high-pressure steam as well as by thermal expansion. The overriding concern in design and operation is to prevent leaks, because of the consequences of the chemical reaction between water and sodium.

Many older fast reactors had evaporators and superheaters in separate vessels, and some had separate sodium-heated reheaters as well. The flow of hot secondary sodium was divided between superheater and reheater and then recombined before flowing to the evaporator. Austenitic stainless steel cannot be used in evaporators because of the risk of chloride stress corrosion. Although the chloride concentration in the feedwater can be controlled by ion-exchange units there is a danger of an accidental increase, particularly if the condenser is cooled with seawater. Evaporator tubes can be made of ferritic steel, such as 2.25 Cr 1 Mo with about 0.4% niobium added to stabilise the carbon, or of a steel with a higher chromium content such as 9 Cr 1 Mo that resists decarburisation. Separate superheaters and reheaters can be made either of austenitic steel (provided they can be kept free from droplets of water from the evaporators) or of a ferritic steel.

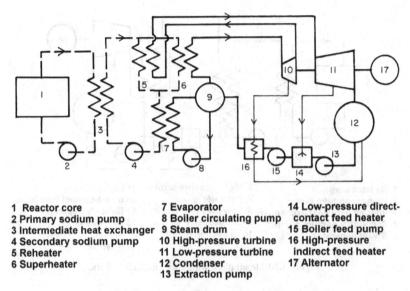

1 Reactor core
2 Primary sodium pump
3 Intermediate heat exchanger
4 Secondary sodium pump
5 Reheater
6 Superheater
7 Evaporator
8 Boiler circulating pump
9 Steam drum
10 High-pressure turbine
11 Low-pressure turbine
12 Condenser
13 Extraction pump
14 Low-pressure direct-
   contact feed heater
15 Boiler feed pump
16 High-pressure
   indirect feed heater
17 Alternator

Figure 4.9 Steam plant with recirculating boilers.

Some reactors used a Lamont-type boiler in which the evaporator produced a mixture of water and steam that were separated in a steam drum, the water being recirculated to the evaporator and the saturated steam being passed to the superheater. Figure 4.9 shows a steam plant with recirculating evaporators and sodium-heated superheaters and reheaters. The main advantage of this arrangement is that the evaporator tube walls are always covered with water. The flow of the two-phase mixture of water and steam is either "bubbly" (near the water inlet and where the steam is in bubbles dispersed throughout the water) or "annular" (near the outlet where the steam bubbles coalesce to form a continuous vapour region in the centre of the tube and most of the water is in a film on the wall). The boiling water on the wall gives a high heat transfer coefficient so that the wall temperature stays close to the saturation temperature. It also has the advantage that the mass of steam and water in the steam drums tends to decouple the reactor and sodium coolant circuits from rapid changes in demand for steam caused by fluctuations in the electrical load, so that control of

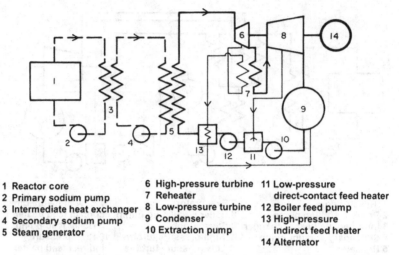

1 Reactor core
2 Primary sodium pump
3 Intermediate heat exchanger
4 Secondary sodium pump
5 Steam generator

6 High-pressure turbine
7 Reheater
8 Low-pressure turbine
9 Condenser
10 Extraction pump

11 Low-pressure
   direct-contact feed heater
12 Boiler feed pump
13 High-pressure
   indirect feed heater
14 Alternator

Figure 4.10 Steam plant with once-through boilers.

the plant is easier. But it has the disadvantage of being complex and expensive.

A "once-through" steam generator as shown in Figure 4.10 is much simpler and cheaper because it does not require steam drum or boiler circulating pumps. The feedwater enters a single heat exchanger in which it is heated to saturation, evaporated and then superheated. The disadvantage is that somewhere along the tube the wall ceases to be covered with water. (This is either the point of "departure from nucleate boiling" (DNB), where nucleate boiling gives way to film boiling, or the "dryout" point, where annular flow gives way to dispersed flow (Collier 1972).) At this point the heat transfer coefficient falls substantially (see section 4.3.3).

It is very difficult to engineer sodium-heated reheaters with a once-through steam generator. The absence of reheat by sodium is a disadvantage, both because it reduces the thermal efficiency of the plant by reducing the mean temperature at which heat is transferred to the steam, and also because the wetness at the low-pressure end of the turbine is increased. There is little that can be done about the reduced efficiency, but the wetness can be reduced by means of moisture separators between the turbine stages or by employing bled-steam reheat as shown in Figure 4.9. Some of the steam is taken from the high-pressure turbine and used to heat the main flow of steam after further expansion. The bled steam is partly condensed in the reheater but is still hot enough to be used in a feed heater. There is a loss of thermal efficiency because of the entropy increase in the reheater, but this may be offset by the increased efficiency of the final turbine stages due to the lower wetness.

4.3.2 Steam Generator Tube Welds

The most important considerations determining the form of the steam generator are the nature of the joints at the ends of the tubes, and the accommodation of differential expansion between the tubes and the shell. If the tubes are welded to tubeplates that form part of the boundary of the secondary sodium circuit the welds are exposed to sodium or sodium vapour, and the frequency of even very small leaks from the steam side must be very low. In some older reactors the tubes were in the form of a U so that the welds at the ends of each tube were positioned above the level of the sodium in the shell and protected from it by a layer of argon. This helped to reduce the thermal stresses at the joints and accommodated relative thermal expansion of the tubes and the shell.

Thermal expansion stresses can be minimised if the tubes are not straight. One such configuration features helical tubes wound in successive layers in an annular cylindrical bundle. This has the advantage that it can accommodate very long individual tubes in reasonably compact units, but the flow pattern on the sodium side is very complex so that it is difficult to ensure uniform heat transfer conditions, and steam generators of this form are expensive to manufacture. An alternative "hockey-stick" design, utilising tubes that are straight over most of their length but with a single bend near one end, has been proposed.

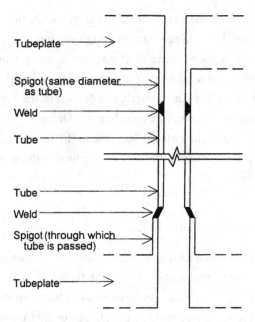

Figure 4.11 Tube-to-tubeplate welds for a straight-tube steam generator.

If straight tubes are used there has to be an expansion bellows in the shell, and care has to be taken in design to ensure that the flow patterns on both sodium and water sides are uniform to minimise temperature differences between the tubes.

In a straight-tube steam generator the tubes can be welded at both ends to spigots machined on the tubeplates. At one end the holes in the tubeplate and the spigots have to be large enough so that, during manufacture, the tubes can be passed through them and welded with an offset weld at one end and a butt weld at the other. The arrangement is shown in Figure 4.11. These welds can be inspected from inside the tubes.

An alternative design that avoids exposing welds to sodium altogether makes use of "thermal sleeves". The tubes pass through individual nozzles in the shell that contains the sodium and are welded to a header positioned outside. The nozzles are in the form of sleeves

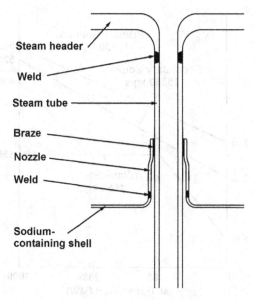

Figure 4.12 A thermal sleeve.

welded to the shell and brazed to the tubes. The brazes and the tube-to-header welds can all be inspected from the outside. The thermal sleeve design has the advantage that the joints between the sleeves and the tubes, and those between sleeves and the shell, are not stressed by the steam pressure, and a leak in neither allows water or steam to come into contact with sodium. The disadvantage is that in a large heat exchanger the manifold, which has to connect many hundreds of tubes, is very complicated and expensive. The arrangement is shown in Figure 4.12.

The problems of steam leaks can be avoided almost completely by the use of double-walled heat exchangers with some means of detecting leakage of either steam or sodium into the space between the walls. The disadvantage is that the heat transfer coefficient is bound to be much worse than in a single-walled heat exchanger, and the cost is higher.

If a single-walled design is chosen the risk of leaks has to be accepted. A leak puts the heat exchanger in which it occurs out of action

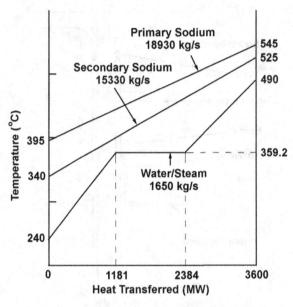

Figure 4.13 Temperatures in the coolant circuits of a 3600 MW (heat) reactor.

until it is repaired. The effect on the availability of the plant can be reduced by having a large number of separate small heat exchangers or heat exchanger modules, any one of which can be shut down for maintenance or repair without reducing the power output of the whole plant by much. On the other hand it is cheaper to build a small number of large heat exchangers. In the end a choice has to be made between better plant availability and lower capital cost. The design provisions that have to be made to ensure the safety of the plant in the event of a leak in a steam generator are described in section 5.3.3.

Useful discussions of steam generator design are given by Hayden (1976) and Lillie (1978) and of the performance of tube-to-tubeplate welds by Broomfield and Smedley (1979).

4.3.3 Steam Generator Heat Transfer

Figure 4.13 shows the temperatures of the coolants of a 3600 MW (heat) reactor with once-through steam generators. Steam is supplied to the turbine at 18.5 MPa and 490 °C at a rate of 1650 kg s^{-1}. The final feed temperature is 240 °C. The total secondary sodium flow rate is 15330 kg s^{-1} and the primary sodium flow rate is 18935 kg s^{-1}.

The choices available in designing the steam generators can be illustrated in a simplified way as follows. Suppose there are N tubes each of length L and diameter D. Then we have

$$Q \simeq \overline{U}\Delta T_m N\pi DL, \qquad (4.4)$$

where ΔT_m is the logarithmic mean temperature difference, Q is the total heat transfer rate, and $\overline{U}$ is the mean heat transfer coefficient which depends on the steam-side conditions. If M is the total mass flow-rate in the evaporators, then $m = M/N$.

A once-through steam generator can be thought of as consisting of three regions as far as heat transfer is concerned: the inlet region where the tube is filled with single-phase water, the boiling region with a two-phase mixture of water and steam, and the superheating region of steam in the vapour phase. The heat transfer coefficient in the superheating region is considerably lower than in the regions where liquid water is present. Values for the heat transfer coefficients for the three regions are given for example by Collier (1972), and can be used to find $\overline{U}$, which depends on m and therefore on N, but not very strongly.

D has to be at least 15–20 mm for ease of manufacture. For a 20 mm tube, $m = 1$ kg s^{-1} makes the mean speed of the two-phase mixture in the centre of the steam generator about 12 m s. If m is much greater than this the problems of vibration and erosion are significant. If D is taken to be about 20 mm the product LN is limited by equation 4.4 and the main choice is between large L (long tubes) or large N (many tubes).

Figure 4.14 shows the sodium, steam/water and tube mid-wall temperatures in a steam generator of the reactor described in Figure 4.13, with $D = 20$ mm and $m = 1$ kg s^{-1}. This choice of m implies that there would be 1650 tubes each about 64 m long. Clearly this rules out straight tubes, but it is reasonable for reasonably compact bundles

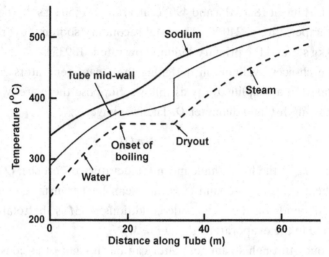

Figure 4.14 Steam generator temperatures (for the plant of Figure 4.13 with 20 mm ID steam tubes).

of helical tubes. An alternative design might select larger diameter tubes. $D = 28$ mm would give $m = 0.5 \, \mathrm{kg \, s^{-1}}$ and $N = 3300$. In this case, allowing for the slightly lower values of $\overline{U}$, the tubes would be 30 m long and could possibly be arranged in six separate straight-tube steam generator units each with 550 tubes.

The discontinuities in the mid-wall temperature shown in Figure 4.14 mark the beginning and end of the two-phase region. Because of the turbulence caused by the formation of vapour heat transfer in this region is better than in the pure liquid region and much better than in the vapour region. An important consequence is a marked step in the tube temperature at the "dry-out" point – the last point where the surface of the tube is wetted. The dry-out point is not fixed but moves up and down the tube both as operating conditions change and at random due to the turbulent nature of the two-phase flow. Thermal stresses are associated with the temperature step, and as they fluctuate there is a possibility of fatigue damage to the tube material. (The extent of these stresses is limited, however, because whatever the heat transfer conditions the tube temperature is constrained to lie between the sodium and steam temperatures that, because of the good heat transfer on the sodium side, are relatively close. In a fossil-fuelled plant where the tube would be heated by a flame at a very much higher temperature the potential is for wider stress fluctuations and therefore greater fatigue damage.)

4.3.4 Plant Efficiency

The steam cycle in Figure 4.9 is very similar to that of a conventional fossil-fuelled power plant, because the maximum steam temperatures are similar. The similarity is to some extent fortuitous, however, because the maximum temperature is set by different considerations in the two cases. In a fossil-fuelled power plant the maximum steam temperature is limited to about 565 °C because anything higher would require the use of austenitic steel rather than ferritic steel, and the increased cost would not be justified by the gain in efficiency. In a sodium-cooled fast reactor plant, however, austenitic steel is widely used, as we have seen. The main temperature limits are the maximum permissible fuel cladding temperature, and the temperature of the structure above the core in a pool reactor or of the hot leg pipework in a loop reactor. The parts of the structure in contact with hot primary sodium are subject to substantial thermal shock if the reactor is shut down suddenly in an emergency, and the primary sodium maximum temperature is limited to a level at which this shock can be withstood.

The problems of withstanding thermal shocks has even led some designers to propose reduction of the primary sodium maximum temperature to below 500 °C (Anderson, 1978; Horst, 1978). Steam superheaters are omitted, and moisture separators and reheaters are incorporated between some of the turbine stages to avoid the irreversibilities associated with high moisture content in the steam. The resulting cycle is very similar to that of a boiling water reactor.

If maximum steam temperatures of 500 °C or above are permissible, a conventional steam cycle with superheat and reheat can be

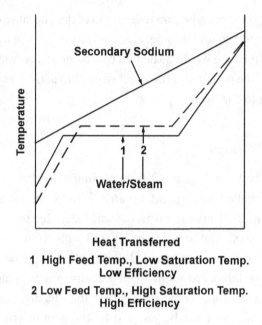

Heat Transferred

**1 High Feed Temp., Low Saturation Temp.
Low Efficiency**

**2 Low Feed Temp., High Saturation Temp.
High Efficiency**

Figure 4.15 The interrelation of plant efficiency with the feedwater temperature.

used, as shown in Figure 4.9. The details of the cycle, and in particular of the feed heating system, may be slightly different from those of a fossil-fuelled power plant because the effect of final feed temperature on efficiency is rather different if heat is being transferred from a relatively low-temperature coolant than if it is transferred from high-temperature gas. The point is discussed in detail by Haywood (1975) and can be illustrated by reference to Figure 4.15.

The plant efficiency is greater the higher the mean temperature at which the working fluid receives heat. Because a large portion of the heat supplied to the fluid is taken up by evaporating it, the higher the saturation temperature, and therefore the pressure, the greater the efficiency. If the cycle efficiency were to be increasedby increasing the steam pressure *without changing the secondary sodium temperatures*, it would be necessary to decrease the feedwater temperature by means of a different feed heating system, as shown in Figure 4.15. This illustrates the fact that the feed and saturation temperatures in a plant of this type are interdependent. In contrast in a fossil-fuelled plant, they are independent (Haywood, 1975).

In this example the increased saturation temperature could, of course, be accommodated by allowing the minimum secondary, and also primary, sodium temperatures to rise. This in turn would require an increase in the secondary and primary sodium flow-rates and would involve the disadvantages of higher coolant speeds, more likelihood of vibration of heat exchanger tubes and fuel elements, greater pressure differences and stresses in the core, bigger circulating pumps and so on.

4.3.5 Available Energy

The use of the secondary sodium circuit to separate steam from radioactive primary sodium carries the disadvantage of a loss of thermal efficiency due to the increase of entropy as heat is transferred from primary to secondary sodium. The extent of the loss can be estimated very conveniently in terms of the available energy (sometimes called the *exergy*). The specific steady-flow available energy a, is given by

$$a = h - T_0 s, \tag{4.5}$$

where h and s are specific enthalpy and entropy respectively, and T_0 is the dead-state or environmental temperature.

If the working fluid in a cyclic power plant circulates at a rate M and receives heat as it changes from state 1 to state 2, the rate at which it receives heat is

$$Q = M(h_1 - h_2) \tag{4.6}$$

and the maximum work output P_{max} is given by

$$P_{max} = M(a_1 - a_2). \tag{4.7}$$

Table 4.1 *Rate of gain of entropy in a fast reactor power plant*

	Rate of gain of entropy (MW/K)
Fission fragments	(0)
Fuel (mean temperature $= 1500\ ^{\circ}$K)	2.40
Primary sodium	4.82
Secondary sodium	5.11
Water/steam	5.82

The maximum thermal efficiency is P_{max}/Q. P_{max} is attained only if all the processes taking place in the plant are reversible. Equation 4.7 shows that P_{max} is the rate of increase of available energy of the working fluid as it receives heat. Every time heat is transferred – from the fuel to the primary coolant, from primary to secondary coolant and from secondary to water and steam – enthalpy is conserved but entropy is increased and therefore the available energy is decreased (equation 4.5).

Table 4.1 shows the steady increase of entropy of the fuel, the primary coolant, the secondary and the steam in a 3600 MW (heat) plant. The entropy of the fuel is greater than that of the source of the energy in the fission fragments, because their kinetic energy has been turned into disorganised thermal agitation (i.e. into heat). The other increases in entropy are due to the transfers of heat from higher to lower temperatures.

Table 4.2 shows the lost potential for doing work represented by each increase in entropy (assuming an environmental temperature of $300\ ^{\circ}$K). The increase in available energy of the steam in the steam generator is 1860 MW, and this would be the net power output if the steam cycle were reversible. There are, however, various irreversibilities – in the turbine for example, due to pressure losses in pipes, and in the feed train – so that the actual net work output would be 1490 MW.

The overall efficiency of the plant is 41%. It is of interest to note that the greatest sources of loss of thermal efficiency – i.e. the greatest

Table 4.2 *Available energy flows in a fast reactor power plant*

Process	Available energy (MW)
Thermal power (K.E. of fission fragments)	3600
Fuel	2880
Primary sodium	2150
Secondary sodium	2060
Water/steam in the steam generators	1860
Electrical output	1490

sources of entropy production – are the change of the kinetic energy of the fission fragments into heat in the fuel, and the transfer of that heat from the fuel to the coolant. In contrast the loss of efficiency due to the presence of the secondary sodium circuit is a mere 0.2%, which is small compared with the losses due to irreversibility within the steam plant. The flow of available energy is shown diagrammatically in Figure 4.16.

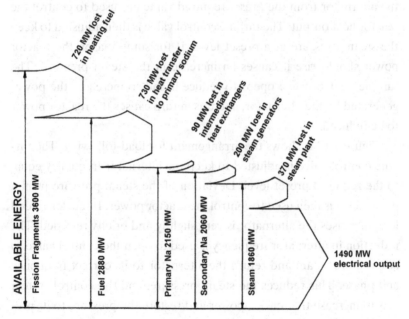

Figure 4.16 Losses of available energy in reactor and power plant.

4.4 CONTROL SYSTEMS

4.4.1 Normal Operation

There are two main strategies for controlling a power plant of any sort, which are chosen according to whether it is required to generate constant power or to respond to a varying demand. The former is adopted for a base-load power plant, and because of their low marginal fuel cost nuclear power stations are often used in this way. If the nuclear generating capacity exceeds the minimum load demanded from the system however some nuclear power stations have to be operated in the latter "load-following" manner.

The appropriate control system for base-load operation is shown in outline in Figure 4.17 A. The "heat source" in the case of a fast reactor power station consists of the reactor together with the primary and secondary sodium circuits. Deviations of the power generated by the alternator from the preset required value are used to control the reactor heat output. The turbine control valve is then adjusted to keep the steam pressure at a preset level. If for some reason the reactor power should rise it causes an increase in the steam pressure. The turbine control valve opens to reduce it and so increases the power generated by the alternator, and this in turn causes the reactor power to be reduced.

Figure 4.17 B shows the arrangement for load-following. The turbine control valve is adjusted to keep the alternator frequency equal to the required preset level. Deviation of the steam pressure from a preset level is then used to control the reactor power. If the demanded load increases the alternator is very slightly and briefly retarded. The reduction in alternator frequency is used to open the control valve to admit more steam and return the alternator to its correct frequency and phase. This reduces the steam pressure, and the control system acts to increase the reactor power and restore the pressure to its preset value. The principles of these control systems are described by Knowles (1976).

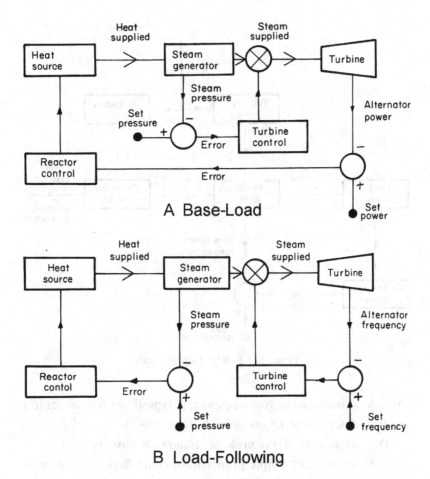

Figure 4.17 Plant control systems for base-load and load-following operation.

Figure 4.18 shows the control of the "heat source" in more detail. One way to adjust the heat output to give the required power output (for base-load) or steam pressure (for load-following) is to use the control variable to control the primary and secondary pump speeds. The reactor control rods are then moved to keep the core outlet temperature at a preset level. If more power is required the primary and secondary pump speeds are increased in step. The increased flow through the core decreases the outlet temperature, and the control rods are

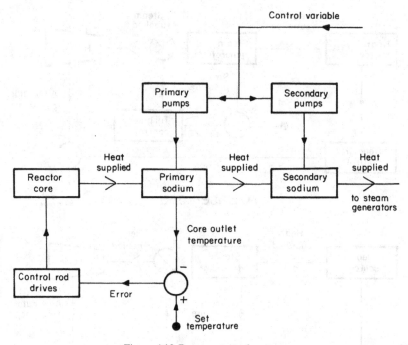

Figure 4.18 Reactor control system.

adjusted to increase the reactor power. A typical fast reactor control system is described by Evans and colleagues (1967).

The various control system elements shown as boxes in Figures 4.17 and 4.18 are not just simple proportional controllers. Derivative or integral terms have to be included in the control function to make the system stable, and the overall gain has to be chosen to keep the main operating parameters within acceptably narrow limits. This is particularly true of the coolant temperatures because fluctuations of the sodium temperature are transmitted so readily to the structure (see section 4.2.4).

In normal operation it is not very difficult to keep temperatures constant. Because of the large mass of sodium in the primary circuit, especially in a pool reactor, power or coolant flow-rate fluctuations cause only very slow temperature changes in most of the structure. The exception may be any structural members, such as control rod guides or mechanisms, exposed to the coolant immediately on leaving the core. Similarly the mass of water in the steam generators, particularly in the steam drums of a recirculating steam generator, and the mass of metal in the drums and heat exchangers, tend to smooth out variations in steam demand so that the steam pressure responds quite slowly to changes in the turbine control valve opening.

There is an important difference, however, between the steam plant of a sodium-cooled fast reactor and that of a fossil-fuelled power station. Because of the lower heat transfer coefficients in a fossil-fuelled station where the heat is transferred to the boiler tubes by convection and radiation from a gas, the boilers are larger than the sodium-heated steam generators of a fast reactor power station. This is not important in normal operation but under "upset" conditions, especially when the steam plant is operating abnormally, the rapid response of the steam generators to pressure changes can create difficulties in control.

4.4.2 Abnormal Conditions

There is a range of abnormal events that befall power stations so frequently that they cannot be thought of as accidents even though they interrupt ordinary operation. It goes without saying that such events cannot be allowed to hazard the operating staff or the public, and in addition the control system has to be designed so that they do not cause damage to the plant.

Examples of these abnormal events are failures of the connection to the power distribution system, which might be due to storm damage to a power line, and would cause a sudden loss of the alternator load. Failure of pieces of auxiliary equipment such as the bearing lubrication system or the alternator cooling system would demand rapid shutdown of the turbine. Failure of the feed pumps or the primary or secondary sodium pumps would prevent full-power operation of the plant and probably require complete shutdown. Perhaps the most important sources of such events are failures of the reactor and plant protective system itself. The protective system is designed to be "fail-safe", so that for example if an ionisation chamber measuring neutron flux goes wrong it gives an indication of high flux. If two or more ionisation chambers fail the protective system takes action as if the flux were really too high and trips the reactor (i.e. shuts it down automatically).

The control system has to be designed so that the plant or the affected part of it can be shut down safely and without damage if one of these abnormal, or "upset", events happens. The design of the turbine and alternator trip system can follow conventional lines, but tripping the reactor and the sodium coolant circuits involves special considerations, the two most important of which are the removal of the decay heat and the avoidance of thermal shock.

Even if the reactor is shut down the fuel continues to generate heat by the decay of radioactive fission products. Emergency cooling systems are provided to remove this heat under accident conditions (section 5.2.4), but it may be preferable not to rely on them in an upset event. It may be better to maintain operation of the primary and secondary sodium circuits in case of a reactor trip and to reject the decay heat to the condenser by maintaining the supply of feedwater to the evaporators. In the event of a turbine trip it may be best to open a bypass so that steam can flow directly to the condenser. If one primary or secondary sodium pump fails it may be decided to trip the reactor but keep the other pumps operating to remove the decay heat.

It is more difficult to avoid thermal shock. If the turbine trips either the full flow of steam has to be diverted to a very large standby "dump" condenser (which is expensive), or the reactor has to be tripped or reduced in power very quickly. If the latter course is taken the sodium flow-rates have to be matched to the power if rapid changes in temperature of the above-core structure and the intermediate heat exchangers are to be avoided.

Considerations such as these impose considerable demands on the control system, especially as the time constants for temperature changes in the reactor core and the steam generators are quite short.

REFERENCES FOR CHAPTER 4

Anderson, C. A. (1978) Optimization of the Westinghouse / Stone and Webster Prototype Large Breeder Reactor, pp 247–259 in *Optimisation of Sodium-Cooled Fast Reactors*, British Nuclear Energy Society, London

Aubert, M., J. M. Chaumont, M. Mougniot, M. Recolin and M. Acket (1978) Temperature Conditions in an LMFBR Power Plant from Primary Sodium to Steam Circuits, pp 305–310 in *Optimisation of Sodium-Cooled Fast Reactors*, British Nuclear Energy Society, London

Broomfield, A. M. and J. A. Smedley (1979) Operating Experience with Tube to Tubeplate Welds in PFR Steam Generators, pp 3–18 in *Welding and Fabrication in Nuclear Industry*, British Nuclear Energy Society, London

Campbell, R. H. (1973) Primary Systems Design of Sodium-Cooled Fast Reactors, *Journal of the British Nuclear Energy Society*, 12, 357–365

Claxton, K. T. (1976) Solubility of Oxygen in Liquid Sodium – Effects on Interpretation of Corrosion Data, pp 407–414 in *Liquid Metal Technology in Energy Production, Volume 1*, American Nuclear Society, Hinsdale, Illinois, USA

Collier, J. G. (1972) *Convective Boiling and Condensation*, McGraw Hill, New York

Eickhoff, K. G., J. Allen and C. Boorman (1967) Engineering Development for Sodium Systems, pp 873–895 in *Fast Breeder Reactors* (BNES Conference proceedings), Pergamon, Oxford

Evans, P. B. F, E. J. Burton, E. Duncombe, D. Harrison, G. O. Jackson and N. T. C. McCaffer (1967) Control and Instrumentation of the Prototype Fast Reactor, pp 765–782 in *Fast Breeder Reactors* (BNES Conference proceedings), Pergamon, Oxford

Frame, A. G., W. G. Hutchinson, J. M. Laithwaite and H. F. Parker (1967) Design of the Prototype Fast Reactor, pp 291–315 in *Fast Breeder Reactors* (BNES Conference proceedings), Pergamon, Oxford

Hans, R. and K. Dumm, (1977) Leak Detection of Steam or Water into Sodium in Steam Generators of LMFBRs, *Atomic Energy Review*, 15, 611–699

Hayden, O. (1976) Design and Construction of Past and Present Steam Generators for the UK Fast Reactors, *Journal of the British Nuclear Energy Society*, 15, 129–145

Haywood, R. W. (1975) *Analysis of Engineering Cycles (Second Edition)*, Pergamon, Oxford

Horst, K. M. (1978) General Electric / Bechtel Prototype Large Breeder Reactor, pp 175–184 in *Optimisation of Sodium-Cooled Fast Reactors*, British Nuclear Energy Society, London

IAEA (2012) *Liquid Metal Coolants for Fast Reactors Cooled by Sodium, Lead, and Lead-Bismuth Eutectic* IAEA Nuclear Energy Series No. NP-T-1.6 International Atomic Energy Agency, Vienna

Knowles, J. B. (1976) Principles of Nuclear Power Station Control, *Journal of the British Nuclear Energy Society*, 15, 225–236

Lewins, J. (1978) *Nuclear Reactor Kinetics and Control* Pergamon, Oxford

Lillie, A. F. (1978) Design of the Clinch River Breeder Reactor Steam Generators, pp 557–571 in *Design, Construction and Operating Experience of Demonstration LMFBRs*, International Atomic Energy Agency, Vienna

Smith, C. A., P. A. Simm and G. Hughes (1979) Analysis of Hydride and Oxide Deposition and Resolution in Relation to Plugging Meter Behaviour, *Nuclear Energy*, 18, 201–214

Tang, Y. S., R. D. Coffield and R. A. Markley (1978) *Thermal Analysis of Liquid-Metal Fast Reactors* American Nuclear Society, Hinsdale, Illinois, USA

Tattersall, J. O., P. R. P. Bell and E. Emerson (1967) Large Commercial Sodium-Cooled Fast Reactors, pp 352–372 in *Fast Breeder Reactors* (BNES Conference proceedings), Pergamon, Oxford

Whittingham, A.C. (1976) An Equilibrium and Kinetic Study of the Liquid Sodium-Hydrogen Reaction and Its Relevance to Sodium-Water Leak Detection in LMFBR Systems, *Journal of Nuclear Materials*, 60, 119–131

5

SAFETY

5.1 INTRODUCTION

5.1.1 Safety and Design

The designer of a fast reactor, just like the designer of any other engineering enterprise, has to take into account what might happen if something goes wrong. He or she has to make sure that whatever happens the risk of injury – either to the operating staff or the general public – or of damage to property, is very slight.

There are basically two ways of making a reactor safe. First the overall design concept is chosen so that it is inherently safe. That is to say that for a number of possible accidents the design is such that the reactor behaves safely and damage does not spread even if no protective action, automatic or deliberate, is taken. But it is not possible to guard against all accidents in this way, however well the overall design is chosen. The second way to make the reactor safe is to incorporate protective systems. These are devices designed specifically to prevent the damaging consequences of accidents. A protective system can be active, such as an automatic shutdown system, or passive, such as a containment barrier.

The design aim is to make sure that the risk to the public is sufficiently small to meet the criteria of acceptability. To ascertain that the aim has been met the designer has to determine the response of

239

the reactor, with its protective systems, to a range of accidents. To test the systems thoroughly it is often necessary to assume that certain accidents happen, even though no way is known by which they could actually take place. These are known as "hypothetical accidents". The final step is to analyse the accidents, whether hypothetical or not, and to ensure that the risks meet the criteria imposed by the authorities that regulate nuclear activities. These criteria vary of course from country to country.

The main concern in reactor safety is to make sure that the radioactive materials – fuel, fission products and activation products – are contained adequately and do not escape to the environment. This is the main subject of this chapter, which is confined to consideration of the safety of the reactor alone. Questions of the safety of fuel manufacture and transport and of waste disposal are not addressed here, nor are other risks associated with the steam and electrical plant, for example.

The safety of nuclear reactors in general is discussed by Lewis (1977) and Farmer (1977). Detailed accounts of some of the subjects touched on briefly here are given by Graham (1971) and Waltar and Reynolds (1981).

5.1.2 Comparison with Thermal Reactors

There are certain features inherent in fast reactors that have important implications for safety because they can affect behaviour in the abnormal circumstances of accidents. The most important is that the fuel of a fast reactor, unlike that of a thermal reactor, is not arranged in its most reactive configuration. This implies that if the integrity of the structure of the reactor core is lost – due to overheating, for example – there is in principle the possibility that the reactivity will increase. The prediction of the extent of the resulting power excursion is discussed in section 5.4. It is necessary to design the reactor system so that the consequences of the excursion can be contained safely.

The severity of a reactivity excursion is determined in part by the mean neutron lifetime, which is much shorter in fast reactors than in thermal. Typical values, as explained earlier in section 1.1.2, are 10^{-3} s for thermal systems and 10^{-6} s for fast.

Both of these differences are disadvantageous in the sense that they tend to lead to greater demands on the safety systems, but against them fast reactors have an important safety advantage in that the power coefficient, of which the Doppler effect is an important component, gives a prompt negative feedback of fuel temperature on reactivity. The importance of the Doppler effect is that whatever happens to the fuel it is a reliable source of negative feedback because it depends on the fuel temperature alone. In a thermal reactor the negative component of the power coefficient depends on the temperatures of the moderator or the structure of the core, and therefore on the transfer of heat from the fuel.

A further striking difference between thermal and fast power-producing reactors is the power density in the core. A peak power density in the fuel of 2.5 GWm^{-3} (see section 3.2.1) implies an average core power density in fast reactors of about 500 MWm^{-3} (averaged over fuel, coolant and structural materials). The corresponding values for thermal reactors are much lower: about 75 MWm^{-3} (PWRs), 50 MWm^{-3} (BWRs) and 3 MWm^{-3} (AGRs).

The high power density in a fast reactor core means that if the fuel is not cooled its temperature rises very quickly. In the extreme case of a hypothetical accident in which a central fuel element were somehow to be deprived completely of its cooling while the reactor was operating its temperature would rise at a rate of some 600 Ks^{-1} and it would melt in 3 or 4 seconds.

If the fuel melts the possibility of rapid vaporisation of the coolant arises. It has been observed that if some molten metals, notably aluminium or steel, are mixed with cold water (as has happened in accidents at foundries, for example) the resulting vaporisation is sometimes violent and a so-called "steam explosion" takes place. The mechanism is not properly understood, and the analogous "sodium vapour explosion" caused by mixing molten oxide fuel with liquid sodium has never been observed to take place with any significant severity. It may indeed be impossible for damage to be done by this mechanism, but until this has been proved the possibility has to be taken into account for sodium-cooled reactors (Board, Hall, and Hall, 1975; Henry and Fauske, 1975).

5.1.3 Low-Pressure Coolants

A major advantage of a liquid-metal-cooled reactor from the point of view of safety is that the coolant pressure is low, so that the primary coolant containment is only modestly stressed and is unlikely to fail, and even if it should fail the coolant does not vaporise. This is in complete contrast to gas- and water-cooled reactors where the coolant pressure is high and extensive protection has to be provided against loss-of-coolant accidents.

It is possible to design a lead-cooled or sodium-cooled reactor so that in the event of a primary circuit rupture the core can be cooled without the provision of emergency supplies of coolant. This can be done by surrounding the vessels and pipework by a leak jacket. Further protection from the consequences of failure of the leak jacket can be provided by surrounding it with a strong concrete structure or by siting the reactor vessel underground. This protects a pool reactor against overheating because the primary coolant cannot fall below the level of the core and the intermediate heat exchangers. Decay heat can be removed indefinitely by the secondary coolant provided the intermediate heat exchangers are intact, or by means of an emergency heat rejection system. In a loop system the coolant pipes have to be connected to the reactor vessel above the level of the core if the core is to remain covered in the event of a pipe break. Decay heat can be rejected if at least one of the primary coolant circuits remains intact.

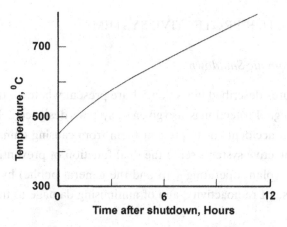

Figure 5.1 Mean primary coolant temperature in a sodium pool reactor after a trip accompanied by complete loss of all secondary cooling.

It is normal to provide auxiliary pump motors to guarantee circulation of the primary coolant at a sufficient rate to remove the decay heat even if all the main pump motors or the electrical supply to them should fail. This may not actually be necessary, however, because natural circulation may be quite adequate to remove decay heat from the core without excessive overheating.

An additional safety feature, particularly in pool reactors, is provided by the large mass of the primary coolant, which may exceed 2000 tonnes. Provided it circulates, by means of pumps or natural convection, its temperature rises only slowly even if there is no secondary cooling at all. This is illustrated by Figure 5.1, which shows the rise in the mean primary coolant temperature for a sodium-cooled reactor, assuming the reactor shuts down and simultaneously all secondary cooling, including the decay-heat rejection system, is lost. There are several hours in which to make secondary cooling available before the temperature rises enough to cause widespread fuel failure (which might happen in the range of 800–1000 °C, depending on the details of the fuel design and the burnup). At atmospheric pressure sodium boils at 892 °C.

5.2 REACTOR PROTECTIVE SYSTEMS

5.2.1 Automatic Shutdown

The features described in section 5.1 are present whatever the details of the design. Protection is also given by systems designed deliberately to prevent accidents or to prevent them from causing damage. Very often protective systems serve the dual function of preventing injury to people (plant operating staff and the general public) by stopping the release of radioactivity, and of minimising damage to the reactor itself.

"Active" protective systems depend on detecting that something is wrong and then taking automatic protective action, which is usually to shut down the reactor. The output from a sensor, such as a thermocouple or a neutron monitor, is amplified and compared with a reference, or "trip", level. If the trip level is exceeded the protective action is taken.

If the trip system is to offer real protection a "fail-safe" system must be employed. This means that if the sensor itself fails the dangerous condition should be indicated. Thus if a thermocouple circuit is broken or short-circuited the amplifier must give an output above the trip level. If the high voltage supply to a neutron detector fails so that it gives zero output the trip circuit must be activated.

It is of course essential to avoid tripping the reactor unnecessarily, and it must certainly not be tripped every time a thermocouple fails. This implies that a "two-out-of-three" system or a variant of it has to be used. Each of the sensors needed for protective action is triplicated. If one indicates danger an alarm is sounded but no other action is taken. If two or more indicate danger the reactor is tripped. This reduces the frequency of spurious trips due to sensor failures because they happen only if two fail at the same time. Introduction of a fourth sensor allows maintenance of one instrument while the reactor is operating without compromising the two-out-of-three reliability. A detailed statistical discussion of the reliability of multiple protective systems is given by Lewis (1977), pp. 103–126.

Different sensors monitor the various reactor parameters that indicate it is operating safely (see section 5.2.2). The output from each two-out-of-three sensor channel feeds in to two or more "guard lines". These are electrical circuits that, when energised, effect the reactor trip – usually by inserting the control rods. The design logic of guard lines is described by Aitken (1977).

To maximise the reliability of the guard lines they are normally redundant, independent and diverse. Redundancy is achieved by providing at least two independent guard lines, either of which is capable of shutting the reactor down. Independence is achieved by ensuring that they are separate, both physically (the components and cables are in different places remote from each other) and electrically (they are supplied from different sources). Diversity is achieved by ensuring that they operate by different principles. For example one guard line may be based on computer software, whereas another may utilise mechanical relays and switches.

The principal action taken automatically when the reactor is tripped is to insert the neutron-absorbing rods – both control rods and shut-off rods – into the core. It is essential that this insertion should be as reliable as possible. A typical arrangement is for the neutron absorber to be connected to the actuator, by which it is moved in normal operation, by an electromagnet. When the reactor is tripped the action of the guard lines is to interrupt the current to the magnet so that the rod falls into the core under gravity.

5.2.2 Whole-Core Instrumentation

The instrumentation needed to detect an incipient accident can be divided into two classes: one for accidents that affect the whole of the core (such as reactivity changes or primary coolant pump failure), and one for accidents that affect initially only part of the core (such as a coolant flow failure in one subassembly). The latter is discussed in section 5.2.3.

It is relatively easy to detect an accident affecting the whole core by means of the instruments used to control the reactor in normal operation. The condition to be guarded against is overheating, so it is necessary to monitor the reactor power and the coolant temperature and flow-rate.

The reactor power is determined by measuring the neutron flux at convenient points. The flux-measuring instruments are normally fission chambers or boron trifluoride chambers. The main difficulty is that neutron flux has to be measured over the range from full power (maybe 3000 MW) to the shutdown level of 100 mW or less – a range of more than 10^{10}. This cannot be done by any single instrument. If the flux is high it is possible to measure the overall ionisation current, but when it is low it is necessary to count individual pulses and determine the average count rate.

Even with these two modes of operation it may not be possible to cover the entire range with a single instrument. It is frequently necessary to have two or more sets of fission chambers in positions of different sensitivity. At low power instruments close to the core are used, whereas at high power other instruments in the neutron shield or outside the reactor vessel are brought into operation.

The high gamma flux from the radioactive primary sodium has to be allowed for. At the highest powers it may be possible to ignore it in comparison with the neutron flux because the energy of a fission event is so much greater than that of a gamma from the sodium. However, at lower powers, when the sodium activity (with a half-life of 15 hours) may correspond to earlier high-power operation, it is necessary to compensate for the ionisation caused by gammas or to discriminate against them if the pulse-counting mode is in use.

Several measuring stations round the core are necessary, partly for reasons of reliability and partly to allow for changes in the flux shape due to the movement of control rods or irregularities in the pattern of loading new fuel into the core. As burnup proceeds the sensitivity of the instruments changes as the flux at the periphery of the core and in the breeder increases relative to that at the core centre. For this reason a high power trip based on neutron flux cannot be very precise, but it is very reliable.

The signal from the neutron flux instrumentation is also used to determine the inverse period dC/Cdt, where C is the indicated flux level. The inverse period is closely (but indirectly) related to the net reactivity, and the reactor is tripped when it becomes too large. This trip system is of little importance when the reactor is operating at power because feedback keeps the net reactivity close to zero, but when the power is very low and very little heat is being generated there is no feedback. Under these conditions an inverse-period trip is a protection against accidental increases in reactivity.

Coolant temperature can be measured by thermocouples at the core outlet. The main difficulty is to ensure that a thermocouple measures the mean temperature correctly, for the coolant temperature is not uniform. Coolant from the edge of a subassembly is cooler than the rest and unless the flow is mixed by some device to enhance the turbulence a thermocouple may be exposed to a stream of unrepresentative coolant; moreover as the coolant flow-rate changes the flow pattern at the outlet may change possibly bringing coolant from a different unrepresentative part of the subassembly to the thermocouple.

Other difficulties are caused by changes in the power generated in a subassembly by the movement of nearby control rods, and by burnup of the fuel. It is thus usually necessary to measure the temperature at a number of positions at the top of the core and use an average for the control and trip systems. The alternative of measuring the temperature farther from the core where the coolant has become more thoroughly mixed is less satisfactory because the delay allows more of the structure to experience a temperature change before corrective action is taken.

Coolant flow-rate can be monitored by flowmeters at the core outlet or by observing the rotation of the circulating pumps. The flow-rate in a pipe can be measured conveniently by an electromagnetic flowmeter, which makes use of the electromotive force induced when a conductor (the sodium) moves through a magnetic field.

It is normal to trip the reactor if the pumps stop or if the flow out of the core falls below a set value. If the reactor is to be operated efficiently at less than its full power, however, either the trip level on coolant flow has to be set low (at say 10% of the full flow-rate) or the trip level has to be altered according to the power required. This disadvantage can be avoided if the ratio of measured power to measured flow-rate is used as a trip signal.

5.2.3 Subassembly Instrumentation

It is more difficult to detect an incipient accident affecting only one subassembly than one affecting the whole core, but it is necessary to do so to prevent the spread of damage (see section 5.4.2). The difficulty arises because there are hundreds of subassemblies in the core. If instruments have to be attached to each, and triplicated, the resulting trip system is very complex. It may also be very cumbersome, especially if instruments have to be fixed to each subassembly after it has been inserted into the core and detached before it is removed.

Thus instruments outside the core that can detect a developing accident anywhere in the core are very desirable. There are two main candidates: failed cladding detectors and boiling detectors.

Failed Fuel. A failed cladding detection system searches for failed fuel elements by monitoring the coolant or cover gas for radioactive fission products that must have come recently from the fuel, and then locates the subassembly in which the failure has occurred. This can be done in two ways. β or γ activity can be sought if the fission products are separated from the coolant, which is already γ-active. Alternatively neutrons from delayed neutron precursors can be sought if a sample of the coolant is removed from the neutron flux.

A typical system has four parts. A sample of the main coolant flow from the core is piped away to a point at which the neutron flux is low. There it passes to a vessel surrounded by moderating material and thermal neutron detectors (usually boron trifluoride counters) that detect any delayed neutron precursors in the coolant. This enables a quick response to cladding failures anywhere in the core, provided the sample has been taken from a point where the flow is well mixed so that it is representative of the flow through the whole core.

Secondly samples are taken from the outlet of each subassembly in turn to a delayed neutron monitor. This serves to locate the sub-assembly in which a failure has occurred. Because there are so many subassemblies to sample, location is relatively slow.

Thirdly a sample of the cover gas over the sodium in the reactor vessel is taken to a γ-spectrometer and a moving-wire β-precipitator. This latter device uses a charged wire to precipitate the daughter products of β-decaying fission products and to transport them to a chamber containing a β-detector. The system thus discriminates against a range of activation products that might be present and selects β-active daughters of gaseous β-active fission products (mainly ^{88}Rb produced by β-decay of ^{88}Kr), so that it responds selectively to cladding failures. Further information about the nuclides present is given by the γ-spectrometer.

Finally the gaseous fission products are stripped from the location system coolant sample by a stream of gas that then goes to a β-precipitator. This provides an independent way of locating failed cladding.

Coolant Boiling. Failed fuel detection has the advantages that it is reliable and a direct indication of the release of radioactivity, which is what has to be avoided. Even a bulk detection system is however relatively slow, taking tens of seconds to detect a failed fuel element, mainly because of the time taken to transport the coolant sample to the detector. It is adequate to control the release of radioactivity to the coolant but a faster system can help to minimise damage to the fuel.

If fuel element failure is caused by overheating it may be accompanied by boiling of the coolant (which takes place at about 920–940 °C at the pressure in the core). A boiling detector gives a quicker indication that something is amiss than waiting for the fuel to fail and then for the failed fuel detection system to operate. Boiling can be detected by acoustic means, as is suggested by the ease with which boiling is heard in a domestic kettle. It is particularly attractive in a reactor because a small number of detectors are enough to detect boiling anywhere in the core.

The main difficulty is that the reactor itself is quite noisy. Sound is generated by the coolant pumps, by the turbulence of the coolant flow, and by cavitation. Cavitation is particularly awkward because it is a form of boiling (caused by local reductions in pressure at points where the flow of the coolant is accelerated, such as at sharp corners or on the blades of the pump impellers), and it makes a very similar noise to boiling caused by overheating. It may be possible to avoid the difficulty by designing the reactor to keep cavitation to a minimum and to discriminate against other background noises by listening in a frequency range in which boiling generates a lot of noise.

As an alternative to acoustic means boiling can be detected by temperature measurement. In some cases boiling in a subassembly can be detected by means of a thermocouple at the outlet, but there are some circumstances in which detection would not be reliable. A partial blockage to the flow somewhere in the subassembly could be large enough to cause severe overheating in its wake, possibly to the boiling point, but at the same time have a very small effect on the total flow-rate and the mean outlet temperature. This is because the resistance of the subassembly to coolant flow is already high and the additional resistance caused by a blockage is small in comparison and reduces the flow-rate only slightly. The effects of subassembly blockages are discussed in more detail later in section 5.4.1.

Table 5.1 *Decay heat produced by radioactive decay of fission products after shutdown from steady reactor operation for an infinite period*

Time after shutdown	Fraction of power before shutdown
1 second	0.062
10 seconds	0.050
100 seconds	0.035
1 hour	0.015
1 day	0.0045
1 week	0.0019
1 month	0.0011
1 year	0.00056
10 years	0.00026

It may be possible, however, to detect a local blockage by observing temperature fluctuations at the subassembly outlet. A partial blockage increases the turbulence of the flow and the differences in temperature between different parts of the flow, and so causes increased temperature fluctuations, or "temperature noise", at the outlet, which can be detected by fast-response thermocouples.

5.2.4 Decay-Heat Removal

So far we have discussed protective systems aimed at detecting an accident in its early stages and preventing it from developing. Another type of protective system mitigates or controls the consequences of an accident. One such is the post-accident heat removal system.

The rate at which "decay heat" is generated by the decay of radioactive fission products in the fuel after a long period of steady operation is shown in Table 5.1 and Figure 5.2. The heat production is slightly less after shorter periods of operation because fewer of the long-lived fission products have accumulated. It will be seen that even a year after shutdown a 3000 MW (heat) reactor generates more than 1 MW.

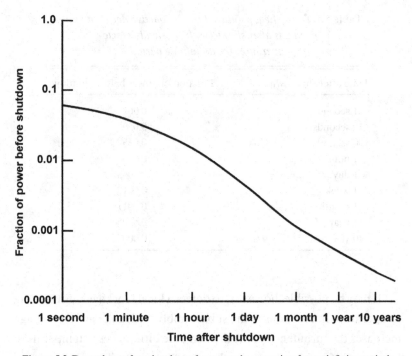

Figure 5.2 Decay heat after shutdown from steady operation for an infinite period.

Because it is essential to remove this heat and keep the fuel cool for a long period after even the most severe of accidents it is necessary to provide several independent and diverse cooling systems.

The first means of removing the decay heat is to use the normal route to the main condenser via the steam generators. However this requires most of the plant to be intact and operable, and in particular the primary and secondary sodium pumps and the boiler feed and condensate extract pumps have to have electrical supplies.

If the steam plant is not available it may be possible to reject heat directly from the secondary sodium circuits, for example by cooling the external surfaces of the steam generator vessels by circulating air. Almost certainly, however, natural convection cannot be relied on and electrical supplies are needed to power the cooling fans, as well as the sodium pumps.

To cater for the possibility that the secondary sodium circuits are not available it is usual to provide separate dedicated decay-heat rejection circuits with heat exchangers in the reactor vessel. The coolant in these circuits rejects heat to the atmosphere in air-cooled heat exchangers. Because sodium freezes at 98 °C for a sodium-cooled reactor the auxiliary coolant can be a mixture of sodium and potassium which freezes below normal atmospheric temperature. Such decay-heat rejection circuits can be designed to operate entirely by natural convection on both the liquid metal and air sides. With such a heat-rejection system the fuel can be cooled safely even if the secondary circuits do not work and there is no electrical power, because natural convection within the reactor vessel can be relied upon to transfer the heat to the decay-heat rejection heat exchangers.

An additional defence against overheating can be provided by cooling the surface of the reactor vessel itself. This can be done for example by surrounding the leak jacket with a water-cooling circuit that rejects heat to the atmosphere and, as in the case of the liquid metal decay-heat rejection circuits, can be made to work by natural convection. A system of this type is sometimes known as an "RVACS" (reactor vessel auxiliary cooling system). Figure 5.3 shows a typical decay-heat rejection system in diagrammatic form.

All these ways of rejecting decay heat depend on the fuel and the structure of the core remaining intact after the accident and on the primary sodium being able to circulate through it. Provisions for cooling the fuel debris after a very severe accident that has destroyed the core structure are discussed in section 5.4.7.

5.2.5 Containment

Radioactive materials have to be kept from being released into the environment both in normal operation of the reactor and in the aftermath of an accident. There are usually three substantial containment boundaries: the cladding of the fuel elements, the primary coolant

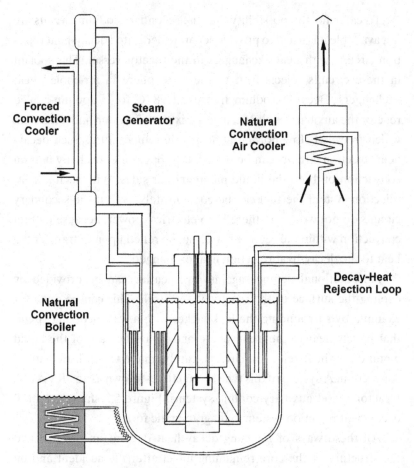

Figure 5.3 Decay-heat rejection systems.

containment (i.e. the reactor vessel in a pool layout, or the reactor, pump and heat exchanger vessels and the connecting pipework in a loop layout), and the reactor building. Fuel and fission products are contained by all three; activation products including ^{24}Na by the last two.

Failure of the fuel cladding is discussed in section 2.4.7. The fuel is designed so that failure of any one element is very unlikely, but there are so many elements in a reactor core (of the order of 10^5 in a 3000 MW (heat) reactor, each being renewed after each year or so of operation at full power) that the possibility of some failures has to be allowed for. Failure by fission-product corrosion or due to a defect in manufacture is likely to result in no more than a small crack in the cladding, which would release to the coolant just the fission-product gases and possibly some volatile fission products. The amount of radioactivity reaching the coolant would be small unless a much larger breach was made in the first place, or the small breach was enlarged by operating the reactor for a long time without replacing the failed fuel. Experiments have shown that even in the event of gross cladding failure the amount of fuel released is very small (Smith et al., 1978).

If there should be widespread cladding failure gaseous fission products would find their way to the cover gas over the coolant. They would be contained by the roof of the reactor vessel, but there might be some leakage through the seals on pump shafts, control rod actuators and rotating shields . The seals have to be designed to keep this, together with the leakage of ^{24}Na, to a low level.

Apart from such minor leaks, the primary coolant containment also has to be designed to prevent a major breach. This means that it has to be able to withstand the effects of accidental loads that might be imposed from without, by a piece of heavy equipment being dropped, for example, or from within by a core accident (see section 5.4.4).

Any release of radioactive material from the primary coolant circuit is contained by the reactor building, or "secondary containment" as it is often known. The building has to have a ventilation system with suitable traps and filters to control any radioactivity released to the atmosphere inside the building and to cope with the effects of the sodium fire that would result from a breach of the primary or secondary coolant circuits. (Ways of preventing such a fire are described later in section 5.3.1.) The building itself has to withstand loads imposed by wind or snow, by earthquakes (if the reactor is to be built in a zone where these occur), by missiles from external sources such as an explosion in a nearby piece of plant or equipment, or by a crashing aircraft.

In summary although care has to be taken in design it is reasonably straightforward to ensure that the three containment boundaries are independent and will remain intact under the loads imposed by a wide range of less severe accidents, which are the ones most likely to occur. This is the principal foundation of the safety of a reactor of this type. Once this foundation has been established release of radioactivity is possible only if there are coincident independent failures of all three boundaries, which is very unlikely, or if some single initiating event can cause breaches in all three.

5.3 OPERATIONAL SAFETY

5.3.1 Operator Dose

Steels are corroded more by water than by liquid metals. As a result the coolant circuits of water-cooled reactors become considerably more contaminated with ^{60}Co and ^{56}Mn than those of sodium- or lead-cooled reactors. This has a marked effect on the radiation doses experienced by the plant operating personnel, particularly the maintenance staff. The principal radiation source in the primary circuit of a sodium-cooled reactor is ^{24}Na, but this decays quickly ($T_{1/2} = 15$ hours) so that a few days after shutdown, provided there have been no fuel failures, radiation levels are low and do not impede access for maintenance. If failed fuel is present the primary circuit cover gas may be contaminated with radioactive Kr and Xe.

Collective doses in the range 10–100 man-mSv per year are reported for sodium-cooled power reactor plants, in contrast to doses in the range 1000–10000 man-mSv per year for BWRs and PWRs.

5.3.2 Sodium Fires

Sodium burns readily in air to produce dense white fumes of sodium peroxide. The reaction is

$$2\,Na + O_2 \rightarrow Na_2O_2 + 11\ MJ \text{ per kg of sodium.}$$

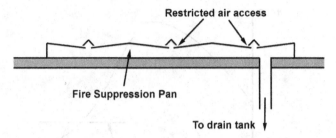

Restricted air access

Fire Suppression Pan

To drain tank

Figure 5.4 Protection against sodium fires.

The peroxide combines with atmospheric water vapour to form sodium hydroxide which is corrosive. The ignition temperature in air is ~250 °C for a quiescent pool but lower, ~125 °C, for a spray such as might emerge from a leaking pipe. Burning sodium can generate temperatures greater than 1000 °C and can cause severe damage to steel structures.

Because the primary sodium is radioactive it is necessary to enclose all the sodium-carrying vessels and pipes in a continuous leak jacket. The space inside the leak jacket is filled with inert gas (nitrogen or argon) and provided with sodium presence detectors. Major secondary circuit components have also to be provided with leak jackets, but they are not necessary for minor secondary components and pipework, which can be located in air-filled spaces with fire-suppression pans on the floors. These pans work by collecting any leaking sodium and directing it to a drain tank while restricting access of air to it so as to minimise the fire, as shown in Figure 5.4.

Sodium pool fires can be extinguished by smothering with common salt (NaCl) or, better, graphite powder which expands by exfoliation to form a blanket over the surface of the sodium.

5.3.3 Sodium-Water Reactions

A defect in a steam tube or a weld in a steam generator of a sodium-cooled reactor may give rise to a leak in the form of a jet of

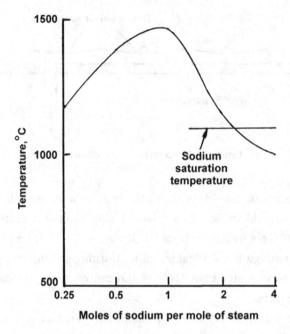

Figure 5.5 Sodium-water reaction temperatures.

high-pressure water or steam into the secondary sodium. The water and sodium react chemically to produce sodium hydroxide, which is strongly alkaline, and hydrogen. The reaction between sodium and liquid water is

$$2\,Na + 2\,H_2O \rightarrow 2\,NaOH + H_2 + 8\,MJ \text{ per kg of water.}$$

Figure 5.5 indicates the temperature of the reaction products as a function of the ratio of the reactants, assuming the reaction is adiabatic and the reactants start at 350 °C with the water as liquid. If the reactants are in the stoichiometric ratio the products reach a temperature of about 1450 °C in steady flow.

Formation of sodium hydroxide is not the whole story because in practice some sodium oxide is formed by the reaction

$$2\,Na + H_2O \rightarrow Na_2O + H_2 + 7\,MJ \text{ per kg of water,}$$

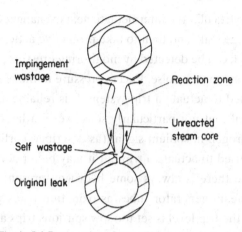

Figure 5.6 Propagation of a small steam generator leak.

but this makes little difference to the temperature of the reaction products.

Small Steam-Generator Leaks. A small hole, such as a crack in a weld or a defect in a steam tube, might be equivalent to a circular hole of diameter 0.1–0.2 mm and give rise to a leak of steam or water into sodium at a rate of 10–100 mg s^{-1}. This would form a hot under-sodium "flame" that could be very damaging. If the "flame" were to play on an adjacent tube it would heat and soften it, the caustic reaction products would corrode it, and the supersonic steam flow would erode the corroded steel and "blow it away", until the tube was penetrated. The water or steam flowing in the tube would act to cool it and delay penetration for tens of seconds or longer depending on the size of the original leak, but eventually the tube would fail and the leak would propagate. Figure 5.6 illustrates the processes taking place as a small leak propagates in this way.

The steam generator has to be equipped with a protective system to minimise the damage caused by a small leak; the system has to prevent the leak from propagating and also prevent the spread of caustic reaction products that might damage the secondary sodium circuit and in particular the intermediate heat exchangers. The system acts to detect the leak, and then to take protective action.

A small leak can be detected by monitoring pressure, the presence of hydrogen, or acoustic noise. Excess pressure in the expansion tank is normally used to actuate a trip system. It is reliable as a means of detecting a leak but not particularly sensitive in a large steam generator. A hydrogen-in-sodium signal (as described earlier in section 4.2.7) can be used to actuate a trip but it may be a poor indicator of a leak because there is always some hydrogen present by diffusion through the steam generator tubes as oxidation takes place on the steam side. If the trip level is set too low spurious trips are unacceptably frequent. Hydrogen leak detection has the additional disadvantage that it is difficult to design a system to respond quickly, in less than 10 s or so. Acoustic leak detection is attractive in principle but in practice it is difficult to discriminate against the noise of the coolant flow and the mechanical rattling of steam tubes against their support grids. Hans and Dumm (1977) survey in considerable detail the various methods of detecting leaks.

On detection of a small leak a trip is initiated to isolate the steam generator on both the steam and sodium sides. The steam is dumped through the safety valves and the intermediate heat exchanger is isolated to prevent contamination with sodium-water reaction products. Operation of the dump system trips both the reactor and the turbine. The isolation and dump system is shown diagrammatically in Figure 5.7. (The bursting discs would not operate in the event of a small leak.)

Once a leak is detected it has to be located and repaired. Acoustic methods may be useful for location, because if one side of the unit is pressurised it may be possible to hear the gas issuing from the leak. An alternative method is to seek sodium hydroxide on the water side of the tubes by chemical means, because it is found that sodium migrates through small leaks against the pressure difference. When the leak has been located the usual method of repair is to plug the affected tube or tubes.

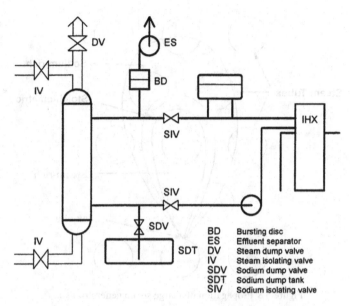

Figure 5.7 Steam generator isolation and dump systems.

In a steam generator made of austenitic steel there is a danger of more extensive damage because of the susceptibility to caustic stress-corrosion cracking. If after manufacture parts of the unit such as the tubeplate welds are left in a state of stress, and there is a leak nearby, the sodium hydroxide formed may cause cracking of the stressed region (Broomfield and Smedley, 1979). This is an important disadvantage of austenitic steels in steam generators and is one of the reasons for the use of ferritic materials.

Large Steam-Generator Leaks. If there is a large leak, such as might be caused by a steam tube breaking in two (an event often known as a "double-ended guillotine failure" or DEGF), water or steam would be ejected into the sodium at a rate of the order of 1 kg s^{-1}. Unless protective action is taken such an event might propagate to failure of other tubes. As in the case of a small leak a reaction "flame" would be formed, but it would be large enough to embrace several tubes. Somewhere in the flame region the reacting mixture would be in the

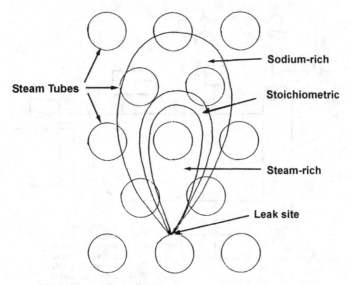

Figure 5.8 Propagation of a large steam generator leak.

stoichiometric ratio above 1300 °C, hot enough to cause the steel of pressurised steam tubes to become soft and to burst. Figure 5.8 illustrates the possible situation.

One kilogram of steam reacting with sodium generates about 1.5 m³ of hydrogen at 1 atmosphere and 350 °C, so in the event of a large leak (1 kg/s or more) large volumes of reaction products are formed very quickly. The secondary sodium circuit is exposed to the steam pressure of 16–20 MPa, possibly higher as the sodium and water react. The protective system has to relieve the pressure quickly. This is usually done by means of bursting discs that release the reaction products to an effluent system that traps the sodium and caustic material in a dump tank and vents the hydrogen and steam to atmosphere. Figure 5.9 shows such a system in outline.

When the bursting discs have opened the secondary sodium circuit is open to the atmosphere, so the intermediate heat exchangers are the only barrier between the radioactive primary sodium and the environment. It is therefore essential that the integrity of the heat exchangers

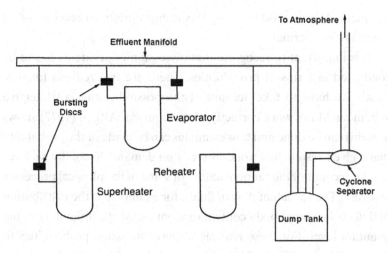

Figure 5.9 Sodium-water reaction-product effluent system.

should not be compromised. The capacity of the effluent system has to be sufficient to protect them from damaging pressures, and the isolation valves have to act quickly enough to prevent contamination by corrosive reaction products. Because of the possibility of propagation it may be necessary to size the bursting discs and effluent system to cope with the simultaneous severance of several steam tubes.

5.4 HYPOTHETICAL ACCIDENTS

5.4.1 Accident Sequences

In the previous sections various inherent features of a fast reactor and various optional protective systems have been identified. Between them these determine what would happen when an accident occurs. To ascertain whether the reactor with its protective systems is acceptably safe we have to follow the development of a series of accidents to find out what would actually happen and what the consequences to the reactor plant, the reactor operating staff and the general public would be. In this way we can determine whether the protection system is adequate. The method of doing this is appropriate to reactors of all types, fast or thermal.

It is important to recognise that discussion of accidents has to be conducted in terms of probabilities. There are two reasons for this. Firstly we have to take account of the possibility that a protective system might not work correctly when required. Aitken (1977) shows how this can be done and how estimates can be made of the probability that a given system will work correctly on demand. Secondly we have to take account of the random nature of some of the physical processes involved. The turbulent flow of fluids, for example, or the distribution of defects in a solid body contain random elements, ultimately at the quantum level. For these reasons we have to assign probabilities to the possible outcomes of events such as the subjection of a certain structural member to a certain load.

Probabilities can best be incorporated in the determination of the outcome of an accident by expressing its development in the form of an "event tree". The method is described by Lewis (1977), pp. 82–84. An event tree is a line that traces the development of the accident. At each point where the development can proceed in either of two ways the line branches and probabilities are attached to each branch. In this way all the possible final outcomes can be identified and the frequency of each determined.

To make a complete assessment of the safety of a reactor in principle we have to construct an event tree for all the possible events that might initiate an accident. This would be impossibly complicated, but in practice it is not necessary to go into so much detail because it quickly becomes clear that the risks to the plant and the public are dominated by a small number of accidents to which attention can be confined.

All accidents are in some sense unpredictable, because if both the nature of the initiating event and the time of its occurrence were known in advance protective measures would be taken and there would be no accident. Many accident-initiating events are quite predictable in nature, however, and unpredictable only in that it is not known when they will happen. Events such as electrical power failure or fuel element cladding failure fall into this category. The resulting accidents can be analysed quite straightforwardly by event trees.

But it is also necessary to guard against initiating events that are unpredictable in nature as well as in timing: to guard against the accident that has not even been thought of. In principle this is very difficult, but in practice it is possible to make sure that all eventualities have been covered. The method is to make pessimistic assumptions in the event trees. For example we can guard against unforeseen failures of a trip mechanism by assuming arbitrarily that it does not work when required, even though we know of no way this could happen. As explained earlier in section 5.1.1 this is called a "hypothetical accident". In the end a judgement has to be made about how much pessimism of this type should be included in the analysis.

Accident sequences for two types of initiating events, those concerning a single subassembly and those involving the whole core, are discussed here. They are representative of the range of hypothetical accidents that have to be analysed to demonstrate the safety of a given design.

5.4.2 Subassembly Accidents

It is pointed out earlier in section 5.2.3 that it is necessary to detect a single subassembly accident and prevent it from spreading to damage the rest of the core. As an illustration of how the protective systems described in section 5.2.3 do this we trace in outline the development of an accident initiated by a partial blockage of the coolant flow in the core region of one subassembly. A complete analysis would be much more detailed but would follow these general lines.

There is of course a very high probability that the protective systems would operate, but the possible sequence of events if they did not is as follows: the blockage would cause overheating, which might

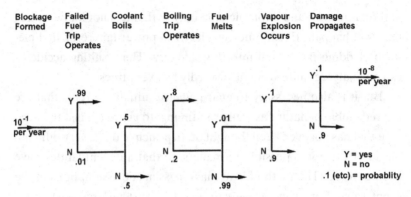

Figure 5.10 An event tree for a subassembly blockage accident.

boil the coolant and then melt the fuel, and this might cause a vapour explosion (see section 5.1.2), which in turn might damage the rest of the core. Figure 5.10 is a very simple form of an event tree describing this sequence. It can be used to determine the frequency with which the undesirable outcome (damage beyond the initially affected sub-assembly) can be expected.

First we have to assess the frequency of the initiating event – the occurrence of the blockage. Subassembly inlets are fitted with filters, so a blockage within the subassembly cannot be formed by material from outside. The only source of blockage material inside the subassembly would be cladding or fuel from one or more failed fuel elements. It might be supposed that fragments of fuel might come out of a crack in the cladding into the coolant and be trapped between the fuel pins to form the blockage. It has been shown that this is very unlikely if not impossible. Experiments in which fuel elements have been made to fail in an operating reactor have shown that large cladding fail-ures have not caused blockages (Kramer et al., 1979; Smith et al., 1978). Data like these show that the frequency of the initiating event is very low.

Since the only source of blockage material is failed fuel a blockage must be accompanied by a substantial failed fuel detection signal. It can be shown that a blockage can cause damage only if it escapes detection either because the failed fuel detection system is inoperative or it happens so quickly that the damage is caused before the detection system has time to initiate a trip (i.e. within a few seconds). Because the detection system may well consist of three separate independent subsystems either possibility is very unlikely, so the probability of failure (the "no" branch at the first branch point in Figure 5.10) is very small.

If the "no" branch is taken, the blockage might become large enough to make the coolant boil. (A more detailed analysis would analyse the growth of a blockage and there would be more branch points in the event tree.) Experiments have determined the temperature distribution in the recirculating wake of a blockage (Kirsch, 1974; Schleisiek, 1974). If the blockage is large enough the coolant will boil ("yes" at the second branch point in Figure 5.10). This is unlikely because the boiling point is some 400 K above the normal coolant temperature, so the blockage has to be quite large.

If the coolant boils it will be detected by the boiling detection system, if one is fitted (see section 5.2.3). Only if this fails can the accident progress. Experiments in reactors (Smith et al., 1978) and test rigs (Schleisiek, 1974) have shown that even if the coolant does boil the fuel is cooled adequately by the two-phase coolant. The blockage has to grow still more before the two-phase flow becomes unstable and the surface of the fuel elements dries out. If this lasts for more than a few seconds (which is unlikely) the affected fuel is uncooled and will melt.

The last two branch points in Figure 5.10 ask whether a vapour explosion takes place and then whether it is violent enough to damage the rest of the core. At present these points have not been resolved experimentally, but the available data (Briggs, Fishlock, and Vaughan, 1979) suggest that the probability of "yes" is low at both branches. It can be shown (Smidt and Schleisiek, 1977) that if a vapour explosion is to be severe enough to disrupt the subassembly wrapper it must involve a large amount of fuel (the amount depends on the design of the wrapper). This in itself makes the probability of "yes" at the last branch point small.

For illustrative purposes the frequency of the initiating event – the formation of a large coolant flow blockage by release of fuel debris from a cladding breach – might be taken to be 10^{-1} per year. The probability that the failed fuel detection system will then fail to initiate a trip might be 10^{-2}. A pessimistic assumption might be that there is then a 50% chance that the coolant will boil within the mass of the blockage or in its wake. A boiling detection system might have a 20% chance of failing. Even if the coolant boils the blockage has to be large before it will cause the fuel to melt, so the probability of fuel melting might be 10^{-2}. If the fuel melts the paucity of experimental evidence of a vapour explosion might be taken to indicate a probability of 10^{-1} that it would occur, and a further probability of 10^{-1} that it would be severe enough to propagate damage through the subassembly wrapper. Thus, as shown in Figure 5.10, the frequency of damage to the rest of the core caused by blockages in single subassemblies would be assessed to be 10^{-8} per year.

It must be emphasised that this is only an illustration of the way an event tree can be used: an analysis for the purposes of a reactor safety case would be based on detailed statistical assessment of the performance of the instrumentation and trip systems and on the available experimental evidence about the various phenomena involved, such as blockage formation, coolant boiling and vapour explosions.

5.4.3 Whole-Core Accidents

Propagation of damage from an event in a single subassembly is only one way in which an accident affecting the whole core might be initiated. There are other ways involving malfunction of the control or cooling systems, the frequency of occurrence of which can be estimated using event trees in a similar way to that described in section 5.4.2.

Table 5.2 *Standard whole-core accident scenarios*

Accident	Description
Slow TOP	Malfunction of one control-rod drive mechanism – one rod withdrawn at the normal operating speed
Fast TOP	Catastrophic failure of the core support – withdrawal of all control and shutdown rods as the core falls away from them with gravitational acceleration
Slow LOF	Simultaneous failure of all primary pump motors – coolant flow reduces as pumps run down and stop
Fast LOF	Catastrophic breach of high-pressure sodium pipes or the diagrid – the diagrid is instantaneously depressurised
LOHS	Instantaneous failure of all the boiler feed pumps.

Notes: "TOP" = transient over-power; "LOF" = loss of flow; "LOHS" = loss of heat sink.

In all cases the frequency is very low because they can take place only if the protective system, which is designed to be very reliable, fails to operate.

To determine the risks posed by this range of hypothetical accidents it is usual to analyse the consequences of a small number of well-defined initiating events, chosen so that they cover the range. Damage would be caused by overheating, overheating would be caused either by excessive reactor power or ineffective cooling, and the extent of damage would depend on the rate at which these happened. By convention a set of standard accident scenarios are analysed. These are described in Table 5.2 and illustrated in Figure 5.11.

(In the event that the single subassembly accident described in section 5.4.2 propagates to the whole core its consequences would be included in the range covered by these standard scenarios. For example if it were to give rise to a vapour explosion the resulting pressure wave might cause widespread cladding failures in other subassemblies, which in turn might cause extensive blockage of the coolant flow. This would not be worse than the Fast LOF standard scenario. Thus the risk posed by single subassembly accidents of this type can be

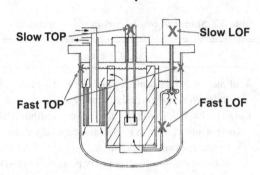

Figure 5.11 Whole-core accident scenarios.

included in the overall assessment of risk by including the frequency of its propagation (10^{-8} per year in Figure 5.10) in the frequency of initiation of Fast LOFs.)

In reality any of these accidents would be terminated safely by the reactor protective system. A trip would be activated by high neutron flux, high core outlet temperature or low coolant flow rate. The trip would act to insert the absorber rods (both control rods and shut-off rods). Redundancy is ensured because only a fraction of the total number of rods would have to be inserted to shut the reactor down and reduce the power to zero. Diversity can be provided by making the absorber rods of different designs, to guard against common-mode failure.

Each of these accident scenarios can in principle be analysed in terms of an event tree, in which the low probability of failure of the various components of the protective system is taken into account at the various branch points. In general the most serious consequences, posing the most severe risks, arise if the protective trip system fails entirely, allowing the reactivity to increase towards prompt critical. This might be the direct result of a Fast TOP, or in the other scenarios it might be caused by ejection of boiling sodium from the centre of the core or compaction of fuel after widespread cladding failure or fuel melting. Accidents involving an increase of reactivity towards prompt critical are usually known as "core-disruptive accidents" or CDAs.

5.4.4 Core-Disruptive Accidents – the Initiation Phase

The increase in reactivity to or close to prompt critical is sometimes, confusingly, called the "initiation phase" of a CDA, in spite of the fact that the accident sequence must have started earlier with an initiating event and the failure of many components of the protective system.

This initiation phase has received close attention since the earliest days of fast reactor development because it addresses the most obvious risk – that because there is excess reactivity available in a fast reactor core it might become prompt critical and suffer a violent power excursion liberating large quantities of energy in a very short time. The fear has been that a fast reactor could "explode", and the object has been to prove that this is impossible and that any actual energy release could be safely contained.

The response time in the prompt-critical regime is related to the prompt neutron lifetime, which in a fast reactor core is of the order of 0.5 μs. Since this is short compared with the time for a sound wave to cross the core there is a possibility that the pressure might rise high enough to generate a shock wave and cause mechanical damage to the containment. This would be analogous to a chemical explosion.

(The mechanism involved differs from that of a nuclear weapon in that the prompt neutron lifetime in the latter is shorter by an order of magnitude or more, so that much larger amounts of energy can be liberated before the core is dispersed. For this reason it is incorrect to characterise fast reactors as "potential bombs". Even with very pessimistic assumptions calculated energy yields for large fast reactor cores do not exceed a few GJ at most, many orders of magnitude less than the yields of the most modest nuclear weapons, which are in the range of 10 TJ or more.)

The first attempts to predict what would happen if a fast reactor were subject to a severe reactivity accident used a drastically simplified model in which feedback reactivity was assumed to be due to thermal expansion of the fuel and therefore to depend linearly on the average fuel temperature (Fuchs, 1946). This approximation was appropriate to very small metal-fuelled cores. It was later realised that in larger cores the fuel could melt and possibly vaporise, and that dispersion of the fuel would be a more important source of feedback (Bethe and Tait, 1956). These early calculations indicated that, for a small core, the energy released by such reactivity transients as could be envisaged was small enough to be contained, but when the same methods were used to assess the effects in larger cores the energy release appeared to be unacceptably large. However as the size of the core increases the neutron energy spectrum becomes softer and the Doppler effect becomes a more important source of feedback, reducing the energy release considerably (Hicks and Menzies 1965). For large power reactors the Doppler coefficient is the parameter that has the greatest effect on the response of the core in the initiation phase of a core-disruptive accident.

Applied to a modern reactor conclusions drawn from the results of these early calculations are not accurate and would not be convincing in a safety case. Nevertheless, because the calculation model is so simple it exposes the essence of the initiation phase and shows clearly if not accurately what would happen. The following illustrations make use of the simple approach introduced by Fuchs.

The main simplifying approximations are to assume that everything happens on the timescale of the prompt neutron lifetime so that it is possible to ignore the delayed neutrons; to represent the state of the reactor core by a single parameter, the average fuel temperature; and to assume that the reactivity feedback depends in a simple way on this temperature. With these assumptions the point-kinetics dynamic equation becomes

$$\frac{dP}{dt} = \frac{(\rho - \beta)P}{\Lambda}, \tag{5.1}$$

and the response of the core is given by an energy equation in the form

$$\frac{dT}{dt} = C(P - P_0). \tag{5.2}$$

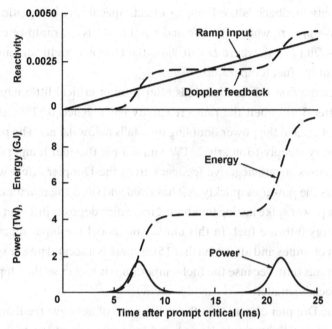

Figure 5.12 The initiation-phase response of a large power reactor to a $50 per second reactivity ramp, ignoring reactivity feedback from dispersion of the fuel.

Here t is time, P is the reactor power and P_0 is the initial power when $t = 0$. ρ is the reactivity, β is the delayed neutron fraction and Λ is the prompt neutron lifetime. T is the average fuel temperature and C is the reciprocal of the heat capacity of the fuel. (It is assumed that in the short time of the transient all the heat is retained in the fuel.)

The reactivity has two components, the input disturbance ρ_i that causes the transient, and ρ_f, the feedback from the fuel temperature. Thus

$$\rho = \rho_i + \rho_f. \tag{5.3}$$

Figure 5.12 shows, according to this idealised model, the response of a large oxide-fuelled power reactor core when reactivity is added at a steady rate of $50 per second, assuming (for the purpose of illustration) that it remains intact and that the only source of negative reactivity feedback is the Doppler effect. Specifically in equation 5.3 we have $\rho_i = rt$, where t is time and $r = 0.15 \, \text{s}^{-1}$ is the ramp-rate; and $\rho_f = -D\ln(T/T_0)$, where $D = 0.008$ is the Doppler coefficient and T_0 is the initial fuel temperature.

For the first few milliseconds after prompt critical little happens but after 5 ms when the ramp reactivity has reached 0.075% above prompt critical the power doubling time falls below 0.5 ms. The power rises very steeply to nearly 2 TW and with it the fuel temperature. This causes strong negative feedback from the Doppler effect which reduces the power as quickly as it has risen and shuts the reactor down after a power spike lasting for about 2 ms, which deposits just over 4 GJ of energy into the fuel. In this unrealistic model the input reactivity ramp continues and after a further 15 ms there is a second power spike, higher this time because the fuel temperature is higher so the Doppler feedback is smaller.

The Doppler effect is not the only source of negative feedback. In addition reactivity would be reduced by dispersion of the fuel. This would be driven by expansion as the fuel temperature rises, especially when it reaches the melting point. After melting the dispersion would be accelerated by the increasing vapour pressure of the molten fuel. Possibly more important would be release of fission-product gas from the fuel crystals (see section 2.3.5). Gas retained in the fuel matrix is in effect stored at very high pressure, and as the crystal structure breaks down this pressure would act on the fuel to disperse it.

Figure 5.13 shows the effect of dispersion of the fuel. The power rises in a sharp spike as before, but in this case subsequent spikes are suppressed by the strong negative feedback as the fuel is displaced outwards.

The energy release of 4.4 GJ shown in Figure 5.13 is typical for a large fast reactor with a Doppler coefficient in the range 0.005–0.010. The implication of calculations of this type is that it is possible to design the reactor containment so that, however severe the pressure transient during the nuclear excursion might be, it can be contained

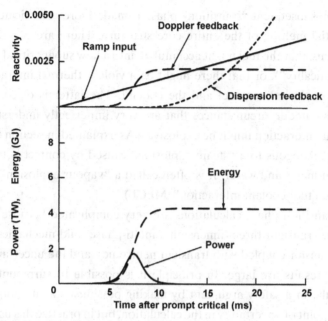

Figure 5.13 As Figure 5.12 but including reactivity feedback from fuel dispersion.

without allowing core material to be dispersed to the environment. It is nevertheless necessary to continue the analysis of the accident to ascertain what happens next, because at the conclusion of the initiation phase the core is left, probably largely molten, containing a large amount of energy and in an unstable state.

5.4.5 Core-Disruptive Accidents – the Transition Phase

During the initiation phase the geometry of the core is reasonably intact. The prompt-critical excursion takes place so rapidly that there is no time for much movement. It is therefore relatively easy to calculate what happens by means of a code (much more complex that the simplified model described earlier) that couples transient neutronics with heat transfer and fluid mechanics, including melting and boiling, in multiple channels.

The subsequent "transition" phase is made more complicated by potential melting of the entire core structure. There are two major concerns: that the fuel might accumulate into a new super critical mass ("recriticality"), or that there might be a violent thermal interaction between the molten fuel and the coolant. The latter is of concern because, under circumstances that are very imperfectly understood, such an interaction might be explosive. (As explained in section 5.1.2, this is analogous to a "steam explosion" caused by contact between molten metal and water. It is often called a "vapour explosion" or a "molten fuel-coolant interaction", MFCI.)

Transition phase calculations are very complicated because they involve transient three-dimensional multi-phase fluid mechanics and heat transfer coupled with transient neutronics, and the uncertainties in the results are large. In principle it is possible to surmount this difficulty in a safety argument by making pessimistic assumptions at every point of uncertainty in the calculation, but in practice this usually results in predicted releases of mechanical energy capable of breaching any reasonable containment.

It is of course impossible to validate a complete transition phase calculation code experimentally without destroying at least one, probably several, complete reactors. Small-scale experiments can however be used to validate individual steps in the calculation. For example kilogram-scale MFCI tests show that sodium vapour explosions are rare and when they do occur they are mild. Similarly small-scale tests on the motion of molten fuel indicate that recriticality in the core is very unlikely.

Figure 5.14 illustrates some of the phenomena and the difficulties encountered in a transition-phase calculation for a Slow LOF accident in a large sodium-cooled core. It shows the state of the fuel, cladding and coolant as functions of axial position and time in one subassembly. As the coolant flow-rate decreases it gets hotter and eventually starts to boil at the core outlet level. The vapour ejects the liquid coolant from the subassembly, mainly upwards into the hot pool. Liquid from

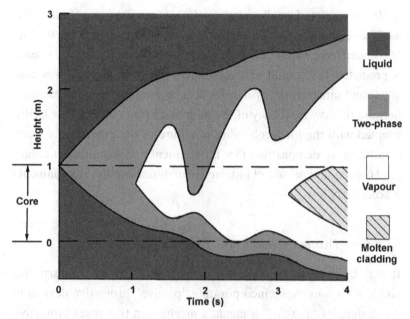

Figure 5.14 Results of a Slow LOF transition-phase calculation showing coolant ejection from and reentry into a subassembly.

the hot pool then falls back into the voided subassembly, boils and is ejected again, and this cycle may take place several times in a few seconds. The motion of the coolant is accompanied by large reactivity changes. When the liquid coolant is ejected the fuel is deprived of most of its cooling and within a second or so the cladding melts so that the fuel itself is free to move. If the fuel remains in the core it then melts after a further period of a few seconds.

The possibility of recriticality arises when the cladding melts. When this happens the fuel pellets or fragments of pellets are likely to be carried out of the core region by the coolant (either vapour or a two-phase boiling mixture) or by fission-product gas escaping from the plena in fuel pins, but there is a small probability that liquid reentering the core may carry fuel towards the core centre. If the fuel melts there is a possibility of an MFCI.

If the results of small-scale experiments are used in a "best estim-ate" (as opposed to "pessimistic") transition-phase calculation the fuel is assumed to be swept out of the core and the resulting energy release is predicted to be mild and containable. Such a conclusion is usu-ally found satisfactory (by nuclear licensing authorities, for example) because the low frequency of the initiating event (the TOP or LOF), coupled with the low probability of failure of the protective system, is sufficient to demonstrate that the frequency of containment failure (and consequent release of radioactive material into the environment) is acceptably low.

5.4.6 Core-Disruptive Accidents – Passive Protection

It may be possible to make the consequences of a core-disruptive accident less severe by incorporating "passive" protective devices in the design. By "passive" is meant a mechanism that takes protective action without external actuation, either by the automatic trip system or by human intervention. There are two main classes: devices to reduce reactivity and devices to prevent recriticality.

Reduction of Reactivity. The reactor trip system works by inserting the control and shut-off rods into the core on receipt of a trip signal (see section 5.2.1). The reduction of reactivity can be made passive by designing the control-rod mechanisms so that the absorbers enter the core in direct response to overheating. This can be done for example by making the core of the electromagnets that attach the absorbers to their actuators of a material with a Curie point above but close to the normal core outlet temperature. In a Slow LOF or Slow TOP accident the outlet temperature rises, the magnets become ineffect-ive and the absorbers fall under gravity into the core. An alternative is to incorporate a component with a high thermal expansion coef-ficient, which responds to the overheating by pushing the absorbers away from the magnets or disengaging mechanical latches, so that they fall.

If the accident has distorted the core the absorbers might not be able to fall freely in their guide tubes. Articulated absorbers, with joints that enable them to negotiate bends, may have a higher probability of entering the core. It may also be possible to increase the chance of insertion by means of spring mechanisms that propels the absorbers downwards when they have been disconnected from the actuators.

Another approach to reducing reactivity automatically when the core is overheated is to increase the neutron leakage. As explained in section 1.6.4, in the case of liquid coolants, loss of coolant from the periphery of the core reduces reactivity because it increases the leakage. The effect can be enhanced in a number of ways. If the neutron reflector above the core or, in the case of a breeder reactor, the upper axial breeder is replaced by coolant (sometimes called a coolant "plenum"), when the outlet temperature rises the density of the coolant falls, leakage is increased and reactivity falls. For a sodium-cooled reactor the effect is much greater in a more severe accident in which the coolant boils and the plenum is filled with vapour.

Radial leakage can be increased by "gas expansion modules", or GEMs as they are often known. These are subassemblies at the periphery of the core that normally contain coolant, but also have reservoirs of trapped gas arranged so that when the gas expands on overheating it expels the coolant. GEMs are attractively simple and reliable, but suffer from the disadvantage that, unless there are very many of them, the amount of reactivity reduction is small. They also have a deleterious effect on the performance of the reactor because they reduce the reactivity in normal operation so that for example the enrichment has to be higher than would otherwise be necessary.

Relocation of Material. The risk of recriticality arises when the fuel becomes free to move, and particularly if it melts. It may be possible to design the structure in such a way that molten fuel is led safely out of the core. One approach is to incorporate in some or all of the subassemblies central ducts that are normally empty except for coolant. The walls of these ducts are made of a material that has a lower melting point than the subassembly wrappers. During the transition phase the molten fuel melts the wall, enters the duct and flows out of the core, either under the influence of gravity or more likely propelled by boiling and vaporising coolant.

A variant of this is to use the control rod and shut-off rod guide tubes, which are already present as ducts through the core. The guide tubes themselves would be made of low-melting-point material and molten fuel could flow out through them. The control rod guide tubes would not be so effective at the beginning of the life of a new core, however, when they would be occupied by the fully inserted control absorbers, but the shut-off rod guide tubes would always be available when the reactor was critical.

The main drawback of any passive mechanism to control the movement of molten fuel is that it would be difficult to demonstrate that it worked correctly. Extensive testing would be required. And it should be noted that such testing of the behaviour of molten fuel as has been done indicates that its "natural" tendency is to disperse, without the provision of any special dispersal path or duct.

5.4.7 Post-Accident Cooling

Best estimate transition-phase calculations, supported by the results of small-scale experiments, indicate that, after a core-disruptive accident, the fuel would be in the form of a mass of debris dispersed in the primary coolant. The internal structure of the reactor vessel (in the case of a pool reactor, the inner vessel, the primary pumps, the intermediate heat exchangers and the decay-heat rejection heat exchangers) would be damaged and probably inoperable but the vessel itself would be intact and would retain the primary coolant. The coolant would continue to serve to remove the decay heat from the fuel and would circulate by natural convection. The coolant itself would lose heat to the emergency cooling equipment, probably the RVACS system (see section 5.2.4). There would be plenty of time to ensure the correct operation of the RVACS (see Figure 5.1).

If the coolant is sodium the fuel debris would fall towards the bottom of the reactor vessel. To eliminate the possibility of it accumulating and forming a critical mass it may be necessary to put in place a structure in the form of a tray shaped to catch the fragments in a sub-critical layer. A tray of this type is usually known as a "core-catcher". It might be made of neutron-absorbing material to reduce the chance of criticality.

Provided it was porous the mass of debris would be cooled by sodium circulating within it. If the fuel were to coagulate in some way so that the sodium could not circulate it might become hot enough to damage the structure on which it rested, so the core-catcher has to be arranged in such a way that the coolant is able to circulate underneath it to keep it cool. The core-catcher would also act to protect the reactor vessel itself from the risk of being damaged and even melted by the hot fuel. In this way the core debris would be retained safely and indefinitely.

REFERENCES FOR CHAPTER 5

Aitken, A. (1977) Quantitative Approach to Reliability of Control and Instrumentation Systems, pp 73–107 in F. R. Farmer (ed.) *Nuclear Reactor Safety*, Academic Press, New York

Bethe, H. A. and J. H. Tait (1956) *An Estimate of the Order of Magnitude of the Explosion when the Core of a Fast Reactor Collapses* Report UKAE-RHM (56) 113, HMSO, London

Board, S. J., R. W. Hall and R. S. Hall (1975) Detonation of Fuel-Coolant Explosions, *Nature*, 254, 319–321

Briggs, A. J., T. P. Fishlock and G. J. Vaughan (1979) A Review of Progress with Assessment of MFCI Phenomenon in Fast Reactors following the CSNI Specialist Meeting in Bournemouth, April 1979, pp 1502–2511 in *Fast Reactor Safety Technology, Volume 3*, American Nuclear Society, LaGrange Park, Illinois, USA

Broomfield, A. M. and J. A. Smedley (1979) Operating Experience with Tube to Tubeplate Welds in PFR Steam Generators, pp 3–18 in *Welding and Fabrication in the Nuclear Industry*, British Nuclear Energy Society, London

Farmer, F. R. (1967) Siting Criteria, a New Approach, pp 303–324 in *Containment and Siting of Nuclear Power Plants*, International Atomic Energy Agency, Vienna

Farmer, F. R. (Ed.) (1977) *Nuclear Reactor Safety*, Academic Press, New York

Fuchs, K. (1946) *Efficiency for a Very Slow Assembly* Report LA-596, US Department of Energy, Washington, DC

Graham, J. (1971) *Fast Reactor Safety*, Academic Press, New York

Hans, R. and K. Dumm (1977) Leak Detection of Steam or Water into Sodium in Steam Generators for LMFBRs, *Atomic Energy Review*, 15, 611–699

Henry, R. E. and H. K. Fauske (1975) *Energetics of Vapour Explosions* Report 75-HT-66, American Nuclear Society, LaGrange Park, Illinois, USA

Hicks, E. P. and D. C. Menzies (1965) Theoretical Studies on the Fast Reactor Maximum Accident, pp 654–670 in Report ANL 7120 *Safety, Fuels and Core Design in Large Fast Power Reactors*, US Department of Energy, Washington, DC

Kinchin, G. H. (1979) Design Criteria, Concepts and Features Important to Safety and Licensing, pp 1–8 in *Fast Reactor Safety Technology, Volume 1*, American Nuclear Society, LaGrange Park, Illinois, USA

Kirsch, D. (1974) Investigations on the Flow and Temperature Distribution Downstream of Local Blockages in Rod Bundle Assemblies, *Nuclear Engineering and Design*, 31, 266–279

Kramer, W. K., K. Schleiseik, L. Schmidt, G. Vanmassenhove and A. Verwimp (1979) In-Pile Experiments "Mol 7C" Related to Pin to Pin Failure Propagation, pp 473–482 in *Fast Reactor Safety Technology, Volume 1*, American Nuclear Society, LaGrange Park, Illinois, USA

Lewis, E. E. (1977) *Nuclear Power Reactor Safety*, Wiley-Interscience, New York

Plein, H. G., R. J. Lipinski, G. A. Carlson and D. W. Varela (1979) Summary of the First Three In-Core PAHR Molten Fuel Pool Experiments, pp 356–359 in *Fast Reactor Safety Technology, Volume 1*, American Nuclear Society, LaGrange Park, Illinois, USA

Schleisiek, K. (1974) *Sodium Experiments Related to Local Coolant Blockages in Test Assemblies Similar to Fuel Assemblies* Report KFK 1914, Karlsruhe Institute of Technology (in German)

Smedley, J. A. (1976) Implications of Small Water Leak Reactions on Sodium-Heated Steam Generator Design, *Journal of the British Nuclear Energy Society*, 15, 153–156

Smidt, D. and K. Schleisiek (1977) Fast Breeder Safety against Propagation of Local Failures, *Nuclear Engineering and Design*, 40, 393–402

Smith, D. C. G., K. Q. Bagley, C. V. Gregory, G. O Leet and D. Tait (1978) DFR Special Experiments, pp 201–214 in *Design, Construction and Operating Experience of Demonstration LMFBRs*, International Atomic Energy Agency, Vienna

Tregonning, K., A. Mackay and K. Buxton (1975) Studies of a Mechanism for Material Wastage by Sodium-Water Reaction Jets, *Journal of the British Nuclear Energy Society*, 14, 77–82

Tucek, K., J. Carlson and H. Wider (2005) Comparison of Sodium and Lead-Cooled Fast Reactors Regarding Severe Safety and Economic Issues Report ICONE13–5037, *Proceedings of the 13th International Conference on Nuclear Engineering*, CNS, Beijing

United States Nuclear Regulatory Commission (1975) *Reactor Safety Study, Appendix 1: Accident Definition and Use of Event Trees* Report WASH 1400, USNRC, Washington, DC

Waltar, A. E. and A. B. Reynolds (1981) *Fast Breeder Reactors*, Pergamon, New York

INDEX

285

Printed in the United States
by Baker & Taylor Publisher Services